Valuing the enviro

Policymakers often face a dilemma: what should they do when the risks of environmental destruction can be averted only by policies that lower standards of living? By setting out a theory of how one can value the environment qualitatively, Raino Malnes offers an answer.

Malnes draws on the tradition of political philosophy to work out public policy guidelines on the basis of careful discussion of the moral dilemmas involved. He tackles directly the problem of setting priorities that reflect the strength of human interests, the extent of environmental degradation, and the need for hard-headed realism. A central issue in the book is the element of uncertainty that often exists in estimates of the environmental effects of human activities. The role of uncertainty is highlighted through a detailed analysis of climate policy and a discussion of environmental dilemmas involving conflict between generations.

Written with great clarity and without technical jargon, this book offers policymakers an alternative to the recent ideas on the quantitative valuation of environmental goods put forward by economists.

Raino Malnes is Professor of Political Science at the University of Oslo, Norway.

Issues in Environmental Politics
series editors Tim O'Riordan, Arild Underdal *and* Albert Weale

already published in the series

Environment and development in Latin America: The politics of sustainability *David Goodman and Michael Redclift editors*

The greening of British party politics *Mike Robinson*

The politics of radioactive waste disposal *Ray Kemp*

The new politics of pollution *Albert Weale*

Animals, politics and morality *Robert Garner*

Realism in Green politics *Helmut Wiesenthal (editor John Ferris)*

Governance by green taxes: making pollution prevention pay
Mikael Skou Andersen

Life on a modern planet: a manifesto for progress *Richard North*

The politics of global atmospheric change *Ian H. Rowlands*

Valuing the environment

Raino Malnes

Manchester University Press
Manchester and New York
Distributed exclusively in the USA and Canada by St Martin's Press

Published by Manchester University Press
Oxford Road, Manchester M13 9NR, UK
and Room 400, 175 Fifth Avenue, New York, NY 10010, USA

Distributed exclusively in the USA and Canada
by St Martin's Press, Inc., 175 Fifth Avenue, New York,
NY 10010, USA

British Library Cataloguing-in-Publication Data
A catalogue record for this book is available from the British Library

Library of Congress Cataloging-in-Publication Data applied for

ISBN 0 7190 4485 5 *hardback*
0 7190 4486 3 *paperback*

First published 1995

00 99 98 97 96 10 9 8 7 6 5 4 3 2 1

Printed by Bell & Bain Ltd, Glasgow

Contents

Acknowledgements

This book is a result of my association with the Energy, Environment and Development Programme at the Fridtjof Nansen Institute. I would like to thank the Director of the Nansen Institute, Willy Østreng, for granting me the opportunity of cooperating so closely with members of his staff.

I have incurred other debts as well. The probing criticism and unwavering friendship of Helge Ole Bergesen have been crucial all along. Warm thanks are also due to two friends who, together with Helge, have alternated as directors of the Energy, Environment and Development Programme over the past five years: Leiv Lunde and Anne Kristin Sydnes. For helpful comments on drafts of greater or smaller parts of the book, I am most grateful to Peder Anker, Per Ariansen, Gunnar Fermann, Andreas Føllesdal, Stein Hansen, Lars Fjell Hansson, Sven Ove Hansson, Jon Hovi, Aanund Hylland, Ingolf Kanestrøm, Richard Purslow, Hilde Nagell, Nils Roll-Hansen, Anne Julie Semb, Henrik Syse, Arild Underdal, Albert Weale, Jon Wetlesen, and an anonymous referee of Manchester University Press. Finally, a special thanks to Anka Nilsen – loving companion.

Introduction

Purpose

This book addresses normative problems that arise in relation to environmental policy. Its aim is to find out what ought to be done if one must choose between (i) allowing an activity that may cause damage to the environment that in turn may erode some people's standard of living, and (ii) not allowing this activity when doing so is a threat to some other people's (possibly one's own) standard of living. Such a situation will be called an *environmental dilemma*. There is no way of resolving it that does not threaten to leave some people worse off than they would otherwise have been.

Two examples of what I have in mind will make it clear that environmental dilemmas number among some of the most pressing problems of environmental policy. First, natural resources may not exist in sufficient quantities to cover both present and future demand. Thus, minerals may be used up at some point in time. Despite recycling and the detection of new deposits, overall supply inevitably declines. Moreover, a renewable resource may be consumed so heavily that its regeneration is wholly or partly undermined. Examples are over-fishing and over-exploitation of arable land. Second, activities that benefit some people may have environmental effects that heighten the vulnerability of others to certain natural strains. It is possible, for example, that the emission of CFC gases (chlorofluorocarbons) destroys the ozone layer; if so, people may in time become incapable of protecting themselves adequately from solar radiation.

As these examples indicate, environmental dilemmas frequently pit the present interests of one group of people against the future interests of another group. A particular problem arises where the two groups belong to different generations. By a generation, I mean all the people alive at a certain point of time when some choice of environmental policy has to be made. If those who have to decide whether or not to engage in a risky activity will be dead when (and if) things go wrong, we have an environmental dilemma involving what may be called *remote risk*. As we shall later see, it is neither obvious nor out of the question that present people – people who are alive today – should ever revoke risky activities out of regard for future people, that is, members of later generations.[1]

The notion of risk figures prominently in the previous elaboration of environmental dilemmas. A risk exists if a certain course of events may, but will not necessarily, cause damage to someone or something. An environmental dilemma involves a choice between two risky actions: either we do something that may damage the environment in a way which may adversely affect someone's standard of living, or we do something else which averts the first risk but instead may leave someone else worse off.

Anyone in the least acquainted with debates about environmental matters knows that the existence and extent of risks of either kind are frequently subjects of controversy. In particular, people who ought to know what they are talking about very often offer divergent estimates of the threat to the environment posed by a certain activity. Theories and data that bear on the issue are contested. This is to say that the normative problem of finding out what to do about an environmental dilemma typically comes in conjunction with another problem: finding out what we are justified in believing about the risks associated with alternative solutions to the dilemma. The latter task, which I shall call *risk diagnosis* (or *knowledge diagnosis*), should of course be taken on first when environmental policy is worked out. It may, after all, prove that an apparent dilemma is spurious – that, for example, there is no real risk of damage to the environment associated with a seemingly risky course of action.

Plan of the book

Chapter 1 will show, using the example of climate policy, how knowledge can be 'diagnosed' so as to arrive at justified beliefs about the risks involved in decisions about environmental policy. Many scientists warn that global mean temperatures will gradually rise to dangerously high levels unless certain activities characteristic of modern, industrial society are not drastically scaled down. We must, they argue, burn less fossil fuels because emissions of carbon dioxide concentrate in the atmosphere and create a greenhouse effect by trapping radiation from the earth. Before we can decide what to do in response to this warning, its credibility must be scrutinized. This analysis serves two purposes, apart from arriving at general lessons for knowledge diagnosis: it contributes to sorting out a political problem that is of tremendous interest in its own right; and it illustrates in detail the nature of environmental dilemmas.

Granted that we face an environmental dilemma (as we arguably do when it comes to deciding on climate policy), what ought to be done? Several normative standards will be proposed. Chapter 2 introduces the *priority principle*, which says, briefly, that we should do what is best for those whose interests are most urgent. It will be justified by laying out a theory about the relative urgency or normative force of various human interests. This is called the *theory of interest*.

Chapter 3 continues the argument in two directions. First, I shall elaborate the priority principle with particular reference to risky choices, that is, situations in which one cannot say for sure whether or not a certain action will cause a setback with respect to someone's ability to fulfil her or his interests. Second, I shall extend the theory of interest to provide guidelines for resolving environmental dilemmas that pit interests of equal urgency against each other. The priority principle says nothing about these cases.

The interest-based standards laid out in chapters 2 and 3 take a purely impersonal approach to adjudicating conflicts between groups who will be differently affected by alternative solutions: priority is accorded the most urgent interest, regardless of whose interest it is. In chapter 4, I turn to countervailing considerations that also ought to play a role in determining the resolution of an environmental dilemma. It is arguable, for example, that in the normal course of events, a person's concern about the prospect that his interests will be fulfilled will exceed his concern about another person's prospect of fulfilling her interests by more than what an impersonal comparison of the relative urgency of the interests in question would justify. People are, briefly put, partial. As I believe normative assessments must take this type of partiality into account, it will sometimes be pertinent to moderate some of the conclusions one may reach in applying the priority principle and other criteria derived from the theory of interest.

Chapter 5 draws on the argument of chapters 2 through to 4 to arrive at guidelines for dealing with dilemmas involving remote risk. The question is: should social and economic adjustments ever be made to protect the interests of people who are yet to be born when those who will bear the brunt of such adjustments are our contemporaries? To answer this question, it must be found what, if any, normative force attaches to the interests of members of future generations. It is often assumed, if only implicitly, that the interests of contemporaries have priority over those of posterity. This idea can be elaborated in different ways. It may imply, for example, that only trivial sacrifices can be demanded of present people for the sake of future generations. I shall argue, however, that the idea is valid only in a much weaker version. The interests of present and future people are on a par *qua* interests, although the ignorance surrounding distant effects of contemporary policies may justify some measure of partiality towards the present, as may the proviso of realism.

In chapter 6, I confront two objections to the perspective on environmental matters adopted in this book. It ends with a final look at climate policy.

The precautionary principle

Those whose interest in environmental matters is mainly political rather than philosophical may question the need for new normative ideas in this area now that the *precautionary principle* is the subject of growing consensus as a sound basis of environmental policy. It is even making its way into legal thinking as a leading candidate for an overall precept of environmental legislation. There are, however, problems about the interpretation of this principle.

A much quoted statement of the precautionary principle can be found in the *Ministerial Declaration on Sustainable Development*, adopted at the international conference "Action for a Common Future" (Bergen, May 1990).

> In order to achieve sustainable development, policies must be based on the precautionary principle. Environmental measures must anticipate, prevent and attack the causes of environmental degradation. Where there are

> threats of serious or irreversible damage, lack of full scientific certainty should not be used as a reason for postponing measures to prevent environmental degradation.

The notions of 'irreversible' and 'serious' damage to the environment are ambiguous. I shall take them to cover at least damage that deprives someone of the means of survival, leaving open whether or not more is intended. What, then, does the precautionary principle prescribe by way of response to a risk of such damage? It tells political authorities to 'anticipate, prevent and attack' it. Again, the language is ambiguous. Literally speaking, to anticipate or prevent a risk is to make sure that it does not arise. If this is what precaution means, it is tantamount to an absolute ban on all activities that have any chance of seriously damaging the environment. We may call this the *strict interpretation* of precaution.

The strict interpretation is contradicted elsewhere. Thus, it is:

> apparent that [the Commission of the European Community] understands 'preventive action' as being different from 'precautionary action.' This is because precautionary action, properly understood, involves some shift in the burden of proof, towards those who would pollute, of demonstrating that pollution is not serious or likely to cause irreversible harm. 'Preventive' is a shade less radical.
>
> (Cameron and Abouchar 1991: 12)

Thus conceived, precautionary action consists of asking those whose activity one suspects of generating a serious environmental risk to allay this suspicion, not of demanding that they immediately eliminate the risk. It is unclear what prevention, being a 'shade less radical', amounts to, but the European Commission definitely does not allude to the literal definition underlying the strict interpretation of the precautionary principle.

Besides the strict interpretation and the European Commission's more lenient one, a third way of defining the precautionary principle is suggested in the proceedings of the Nordic Council's "International Conference on the Pollution of the Seas" (October 1989). It says that there is a:

> need for an effective precautionary approach, with that important principle intended to safeguard the marine ecosystem by, amongst other things, eliminating and preventing pollution emissions where there is reason to believe that damage or harmful effects are likely to be caused,

> even where there is inadequate or inconclusive evidence to prove a causal link between emissions and effects.
>
> (Quoted from Cameron and Abouchar 1991: 16)

In juxtaposing 'elimination' and 'prevention', the Nordic Council seems inclined to a literal reading of the latter. Yet, it recommends preventive measures only in case 'damage or harmful effects are *likely* to be caused' (italics added). Some environmental risks are, accordingly, too low to call for action, but there is no saying where the line between likely and unlikely damage goes. This may be called the *probabilistic interpretation* of the precautionary principle.

Yet another interpretation of the precautionary principle is found in the "Convention for the Prevention of Marine Pollution by Dumping from Ships and Aircraft" – the so-called Oslo Convention – that entered into force in 1974. It says that:

> Contracting Parties pledge themselves to take all possible steps to prevent pollution of the sea by substances that are liable to create hazards to human health, to harm living resources and marine life, to damage amenities or to interfere with other legitimate uses of the sea.
>
> (Quoted from Hey 1991: 249)

The ambiguous phrase is '*liable* to create hazards' (italics added). It means either 'capable of creating hazard', which points towards the strict interpretation of the precautionary principle, or 'having a strong tendency to cause hazard', which has connotations of the Nordic Council's probabilistic interpretation.

To conclude, the precautionary principle does not clearly define the steps to be taken in case precaution is called for. In one interpretation, it approximates to a blanket prohibition of risky activity; in another, it signals permissiveness in the face of environmental risk. Even more problematic, however, is the fact that the precautionary principle reckons with only one side of the account – the cost of environmental destruction – and pays no attention to possible costs of preservationist measures. As indicated earlier, and as chapter 1 will show in detail, the elimination of environmental risk may be a risk in its own right.

Thus, the various statements of the precautionary principle point in different directions, and none comes near to being a determinate guideline for environmental policy. But the general thrust of the principle is obviously sound: there may be risks of environmental destruction that we should not permit ourselves to live with. This book works out general guidelines for identifying such risks through discussions of moral considerations that bear on the relationship between humankind and nature. Thus, it may be seen as an effort to render the politically prominent precautionary principle more precise and clarify its rationale.

Note

1 For the sake of convenience, we may think of 'present' and 'future' people as referring to generations that are completely removed, having no members whose lifetimes overlap. A discussion of related conceptual problems is found in Laslett and Fishkin (1992: 6–11).

1

Environmental dilemmas: the case of climate policy

Introduction

There are many loud warnings that emissions of carbon dioxide, if uncurbed, will lead to global warming, which in turn will precipitate climate change during the next century. Such change, it is furthermore argued, will threaten living conditions in large parts of the world. There are also, however, some who warn that curbing emissions of carbon dioxide to the extent allegedly needed to stabilize the global climate may seriously undermine economic growth, especially in developing countries. Are we, then, caught on the horns of an environmental dilemma? Do we face the choice between two options that both place someone's standard of living in jeopardy?

Either warning alluded to above cites a *risk* that something undesirable will result from a certain course of action, but neither of these risks is undisputed. Some analysts see no real danger in unabated emissions of carbon dioxide, while others deny that abatement is a threat to economic growth. Thus, serious disagreement surrounds the risks associated with both responses to the warning of climate change. This kind of situation is well known as far as environmental matters are concerned. It raises the question: how are we to distinguish between true and spurious risk? This chapter is an attempt to ascertain how this can be done. It discusses several ways of arriving at justified beliefs about the nature and extent of alleged risks. This is best done in relation to concrete issues, and I will take the scientific debate about climate change as the point of departure.

No cases of environmental policy making are alike, but the following discussion still yields general insights about ways and means of diagnosing knowledge in this field. Two caveats are, however, in order. In the first place, the climate issue is immensely complex, encompassing a number of subjects whose analysis involves contributions of many sciences, from chemistry to sociology. The subsequent argument will predictably be deficient with respect to empirical matters. Not only is my knowledge of existing research in this field scant, but new results are constantly being published, so that no account of climate change can long remain authoritative. But this need not be a source of serious concern. The chapter serves its purpose provided it gets the basic features of the climate issue roughly correct and points towards sound guidelines for diagnosing knowledge about risk.

In the second place, some readers may find the present chapter long and tedious. Yet, a fairly detailed account of the climate debate is needed to arrive at general guidelines for risk analysis. It should be mentioned, however, that the subsequent discussion of normative standards for resolving environmental dilemmas (chapters 2–5) can be read independently of what the present chapter says about ways of ascertaining whether or not such a dilemma exists.

The greenhouse theory

It has long been known that heat-trapping gases concentrate in the atmosphere and block outgoing radiation from the earth. Their presence makes the earth much warmer than it would otherwise have been – so much so, in fact, that it would have been uninhabitable in their absence: these gases raise the average global temperature by about 33°C. By far the most important gas is water vapour, but warming is also precipitated by atmospheric concentrations of carbon dioxide, chlorofluorocarbons (CFCs), methane, and nitrous oxide. The contribution of these so-called *greenhouse gases* to making the earth a warmer place is the *greenhouse effect.*

A major component of the greenhouse effect is natural in the sense that human activity plays no part in generating it. A minor component, which is called the *enhanced* greenhouse effect, stems from activities that are characteristic of industrial society, such as the use of fossil fuels. This effect spurs a growing concern that an unprecedented increase of the earth's average temperature can be traced to industrial emissions of greenhouse gases. The main culprit is carbon dioxide – a by-product of the burning of fossil fuels in electricity production, transportation and innumerable industrial processes. The concentration of carbon dioxide in the atmosphere is estimated to be about 25% higher today than it was before the Industrial Revolution. Some scientists and environmentalists believe that this and other contributions to the enhanced greenhouse effect – such as rising methane concentration caused by rice cultivation, cattle rearing and biomass burning – can have disastrous consequences. According to Michael Grubb (1989: 3),

> if emissions of greenhouse gases continue to grow as widely expected, the global average temperature by the latter half of the next century will have increased by an amount which is comparable with the warming since the depths of the last ice age to the present.

In Daniel Kevles' (1989: 33) view,

> when the level of carbon dioxide and equivalent trace gases doubles from preindustrial concentrations – which will happen some time during the next century, at current emission rates – the average global temperature will rise eight degrees Fahrenheit; and ... yield a climatological shift of drastic proportions.

The scientific basis of the claim that emissions of heat-trapping gases are in the process of precipitating a drastic climatological shift is the *greenhouse theory*. Its most prominent and authoritative statement comes from the International Panel on Climate Change (IPCC). The IPCC is a research group composed of scientists from more than 60 countries, formed in 1988 at the initiative of the United Nations Environment Programme and the World Meteorological Organization. Its tasks are to assess 'the scientific information that is related to the various components of the climate change issue, such as emissions of major greenhouse gases and modification of the Earth's radiation balance resulting therefrom,' and to formulate 'realistic response strategies for the management of the climate change issue' (IPCC 1990: Preface). In 1990, the group published a three-volume report: the first volume is an analysis of the risk that anthropogenic greenhouse gas emissions will cause climate change, the second an assessment of the environmental and social effects that may follow from such change, and the third a discussion of response strategies aimed at forestalling it. The IPCC report was subsequently endorsed by the Second World Climate Conference (Geneva, November 1990) and the United Nations General Assembly (December 1990) as a scientific basis for international negotiations on limits to future emissions of greenhouse gases. Such negotiations were commenced in February 1991. Meanwhile, the IPCC continued its work and in 1992 published a supplement to the original report (IPCC 1992), revising some premises but reiterating its major conclusions.

The greenhouse theory has three components. It starts from a claim to the effect that the global mean temperature has increased over the last 100 years owing to growing atmospheric concentrations of anthropogenic greenhouse gases. Its second component, which will be referred to hereafter as the *greenhouse prognosis*, adds that further emissions of these gases at present rates will cause the temperature to increase so much that the global climate will change markedly. And the third component is the prediction that climate change of the kind in question will disrupt living conditions in large parts of the world through its adverse effects on the environment. I shall elaborate the three components of the greenhouse theory in due order.

The greenhouse explanation of past temperature trends

According to the IPCC, 'Global-mean surface air temperature has increased by 0.3°C to 0.6°C over the last 100 years, with the five global-average warmest years being in the 1980s,' although the 'increases have not been smooth with time, nor uniform over the globe' (IPCC 1990: xii). In this trend, the IPCC sees evidence of an enhanced greenhouse effect. It argues, more specifically, that the observed increase in global mean surface air temperature can best be explained as an effect of growing atmospheric concentrations of anthropogenic greenhouse gases. As mentioned above, carbon dioxide is the prime culprit. The increase in its atmospheric concentration during the last 150 years, mostly traceable to the burning of fossil fuels, is said to account 'for over half the enhanced greenhouse effect in the past' (*ibid.*: xi).

The IPCC arrives at this explanation by comparing past climate trends with temperature predictions elicited from *general circulation models* (GCMs). These models purport to depict the atmospheric effects of greenhouse gas emissions. By synthesizing 'our knowledge of the physical and dynamical processes in the overall [climate] system and ... the complex interactions between the various components' (*ibid.*: xx), they permit calculations of how much mean temperatures will rise or fall after a certain change in the concentration of greenhouse gases. Feeding the models with historical data about emissions of carbon dioxide and other heat-trapping gases, one arrives at *post hoc* predictions which, according to the IPCC, fit the observed temperature increase over the last 100 years quite well.

Needless to say, however, the fact that a scientific model yields accurate predictions is no decisive test of its validity. An empirical pattern can always be explained in more than one way, and, in view of this, it matters greatly whether or not a given model is constructed on the basis of comprehensive, well established knowledge of the phenomenon it allegedly represents. If we feel certain that no important causal factor has been left out and that we have a firm grasp of how causal mechanisms work, predictive success means a lot more when it comes to validating the model than it does if we have before us a tentative effort at depicting unfamiliar structures and processes.

How do the GCMs fare in this respect? The IPCC avows that their level of theoretical sophistication is low: 'in their current state of development, the descriptions of many of the processes involved are comparatively crude' (*ibid.*: xx). A particular problem lies in modelling *indirect* greenhouse effects, that is, interactions in the atmosphere that are set off by increasing concentrations of greenhouse gases and that either reinforce or counteract the effect on the climate of such concentrations (Isaksen 1992: 2). The most conspicuous deficiency of the GCMs is their inability to account for how atmospheric concentrations of water vapour – the most important natural greenhouse gas – vary with changes in the concentration of anthropogenic gases. This phenomenon is sometimes invoked to explain the remarkable stability of the earth's climate. If the air warms, the level of moisture increases and more clouds form to keep sunlight out, causing a cooling of the climate. If the climate becomes cooler and drier, clouds become fewer and let more solar radiation in to warm the earth. The process works both ways to stabilize temperature within fairly narrow limits. According to one view, the threat of global warming may be neutralized by the first loop of this feedback mechanism. While a greater concentration of carbon dioxide in the atmosphere traps radiation and pushes up the temperature, it will subsequently fall because more warmth brings more clouds that shield the earth. Another view, endorsed by the IPCC, says that the feedback mechanism is positive in the sense that climate change caused by anthropogenic gases will be reinforced by water vapour (Isaksen and Fuglestvedt 1993: 4).[1] Anyway, the reaction of water vapour to added concentrations of anthropogenic gases has yet to be incorporated in climate models, and its incorporation is a daunting task.[2]

The IPCC (1990: xvii) still pronounces 'overall' and 'substantial' confidence in GCMs, believing them to 'predict at least the broad-scale features of climate change' (*ibid*.: xxvii). It grounds this confidence in their 'generally realistic' simulation of the present climate, their 'generally satisfactory portrayal of aspects of variability of the atmosphere,' and their ability 'to simulate important aspects of the climate of the last ice age' (*ibid*.: xxviii). But the guarded language and some explicit reservations about errors and omissions bespeak considerable uncertainty as regards the use of GCMs for explanatory purposes.

As mentioned above, a supplement to the IPCC report of 1990 was published in 1992. According to the authors, the new assessment prompts no major revision of the greenhouse theory. 'Findings of scientific research since 1990 do not affect our fundamental understanding of the science of the greenhouse effect and either confirm or do not justify alteration of the major conclusions of the first IPCC Scientific Assessment' (IPCC 1992: 5). It turns out, however, that some premises have been amended. One factor that did not explicitly enter into the original explanation of past climate trends is the cooling effect of tropospheric and stratospheric concentrations of aerosol compounds. These are composed of sulphates – airborne particles of about 0.1 to 1 micron in diameter, mostly emitted from industrial activity and largely concentrated over industrialized areas in the northern hemisphere. They absorb solar radiation and reflect it away from the earth. The IPCC (*ibid*.: 19–20) estimates that,

> [f]or clear-sky conditions alone, the cooling caused by current rates of emissions [of sulphur compounds from anthropogenic sources] has been ... about 1Wm^{-2} averaged over the Northern Hemisphere, a value which should be compared with the estimate of 2.5Wm^{-2} for the heating due to

> anthropogenic greenhouse gas emissions up to the present.... At the time of [the first IPCC report] ... it was recognized that sulphate aerosols exert a significant negative radiative forcing on the climate but this forcing was not well quantified. Since then progress has been made in understanding radiative forcing by sulphate aerosols, and an additional source of negative forcing has been identified in the depletion of stratospheric ozone due to halocarbons.
>
> (IPCC 1992: 22)

The IPCC believes, however, that the inclusion of both sulphate emissions and decreasing ozone concentrations strengthens rather than invalidates the greenhouse explanation of past temperature trends. Aerosols have, as indicated above, counteracted the heating effect of anthropogenic greenhouse gases, and because increases in stratospheric ozone contribute to the enhanced greenhouse effect, declining ozone concentrations reduce it. Thus, the 'lack of these negative forcing factors in GCMs does not negate the results obtained from them so far' (*ibid.*: 22). Indeed, the

> consistency between global temperature changes over the past century and model simulations of the warming due to greenhouse gases over the same period is improved if allowance is made for the increasing evidence of a cooling effect due to sulphate aerosols and stratospheric ozone depletion.
>
> (*Ibid.*: 7).

The greenhouse prognosis

The greenhouse prognosis is derived from GCMs. The IPCC (1990: xi) predicts, more specifically, that based on 'current model results', continuing global emissions of anthropogenic greenhouse gases at 1990 levels will bring 'a rate of increase of global mean temperature during the next century of about 0.3°C per decade (with an uncertainty range of 0.2°C to 0.5°C per decade)'. The upshot is 'a likely increase in global mean temperatures of about 1°C above the present value by 2025 and 3°C before the end of the next century' (*ibid.*), which will lead to significant climate change, expected to be triggered by a rate of heating somewhere between 0.2°C and 0.5°C per decade.

The greenhouse prognosis is premised on quantitative estimates of future anthropogenic emissions of gases such as carbon dioxide and methane. The IPCC starts, more specifically, from an 'emission scenario' called 'Business-as-Usual' (BaU), in which, as the name indicates, 'few or no steps are taken to limit greenhouse gas emissions' (*ibid.*: xviii). This is to say that the current 2% increase in anthropogenic carbon dioxide in the atmosphere will continue, doubling the total carbon dioxide equivalent of anthropogenic gases before the middle of the next century. The underlying assumptions – the building blocks of BaU – are as follows: world population approaches 10.5 billion before the end of the next century; economies grow at a rate of 2–5% in the coming decade but stall thereafter; energy supply is coal intensive; energy efficiency shows little progress; carbon dioxide controls are modest; tropical forests are depleted; and nothing is done to curb emissions of methane and nitrous dioxide from agricultural production (*ibid.*: 341). In short, BaU extrapolates demographic, social and economic developments that have been characteristic of industrial societies in the present century.

The *1992 IPCC Supplement* reconsiders some elements of the greenhouse prognosis. First, it adjusts the estimate of future anthropogenic emission of methane, arguing that it will be 'significantly lower ... than reported in IPCC 1990' (IPCC 1992: 12). Second, the BaU scenario is supplemented with six alternative scenarios embodying 'a wide array of assumptions ... affecting how future greenhouse gas emissions might evolve in the absence of climate policies beyond those already adopted' (*ibid.*: 13). No scenario is assigned a greater likelihood than others; it is a question of possible and widely diverging futures, some involving far higher carbon dioxide concentrations than the BaU scenario. Thus, a scenario presupposing 'moderate population growth, high economic growth, high fossil fuel availability and eventual hypothetical phaseout of nuclear power' inflates BaU projections by around 25% (*ibid.*: 15). Other scenarios, however, involve low economic growth, dwindling supplies of fossil fuel and carbon dioxide emission paths that lie below that of BaU. But the IPCC sets little store by the latter. 'Overall,' it concludes, the enlarged list of scenarios indicates 'that greenhouse gas emissions could rise substantially over the coming century' (*ibid.*). Hence, there is no reason to revise the original estimate of future greenhouse-induced warming rates (*ibid.*: 24–5).

The environmental and social effects of climate change

The United States Environmental Protection Agency wrote in a report to Congress in 1988 that 'climate change could result in a world that is significantly different from the one that exists today' (quoted in Colglazier 1991: 64). The envisaged change is twofold. First, climate change risks upsetting biological and physical processes and interfering with fine-tuned ecosystems. It may, in particular, lead to increasing evaporation from water basins, leaving dwindling freshwater supplies in many places. This falls under what we may call environmental effects of climate change. Second, environmental effects may bring adverse social effects in their wake. Scarcity of freshwater will, for example, have serious implications for human wellbeing.

The IPCC concentrates on three environmental effects of climate change. The first is rising sea levels. If the oceans warm, and glaciers as well as the polar icecaps and ice floats melt, sea levels may surge to such an extent that coastlands are flooded and people and wildlife driven inland. The BaU emission scenario is believed to bring about 'an average rate of global mean sea level rise of about 6 cm per decade over the next century (with an uncertainty range of 3–10 cm per decade), mainly due to thermal expansion of the oceans and the melting of some land ice' (IPCC 1990: xi). The aggregate increase will be about 20 cm during the first half of the next century, approximating to 65 cm above the present level before 2100. Given the facts that one-tenth of world's population lives within 20 km of the coast and population density of this coastal zone exceeds 100 people per km^2 (Hayes 1993: 129), the predicted increase may have very adverse social effects.

Needless to say, this prediction is uncertain. There are many sources of error in estimates of how much oceans will expand through flows of runoff water from melting mountain glaciers and the Antarctic icecap. Even analysts who generally endorse the greenhouse theory stress this problem. Thus, Richard A. Warrick and Atiq A. Rahman (1992) place the best estimate of mean sea level rise under BaU at 60 cm, with an uncertainty range of ±45 cm. It is difficult, moreover, to foretell the effects of rising sea levels on human life conditions. They are apt to vary from region to region, depending on topographic factors and the extent to which coastlines can be protected. Still, proponents of the greenhouse theory conjure up some frightening prospects: coastlands will be flooded, 'people and wildlife ... driven inland', and 'salty oceans ... surge upstream through the mouths of the rivers' (Kevles 1989: 33–4). They also argue that the most vulnerable areas are found in some of today's poorest countries, which cannot invest much in the protection of coastal structures.[3]

Another environmental effect of climate change lies in disturbances of complex ecological systems. Some fear that rising temperatures will bring higher moisture levels, which will wipe out plants whose conditions of growth are already delicate. This risk is particularly pronounced with respect to tropical forests, which 'alone support about 200 million people whose survival and self-conception depend on the ecological stability of those forests' (Brown 1992: 213). To be sure, plant species will respond differently to a warmer climate, some thriving while others become extinct, and there are no models available to forecast local changes in precipitation and soil moisture (IPCC 1990: xxx–xxxi). This implies, for example, that effects on the ecological conditions of agricultural production are uncertain. A review of recent research concludes: 'On balance, the evidence is that food production at the global level can, in the face of estimated changes of climate, be sustained at levels that would occur without a change of climate, but the cost of achieving this is unclear. It could be very large' (Parry and Swaminathan 1992: 121).

Rising global mean temperatures may also beget a more turbulent climate that involves higher frequencies of days with extreme weather. Again, however, scientific projections are uncertain. On the one hand, the IPCC (1990: xxiii) sees 'no clear evidence that weather variability will change in the future'; on the other hand, 'with a modest increase in the mean, the number of days with temperatures above a given value at the high end of the distribution will increase substantially,' which is to say that 'the number of very hot days or frosty nights can be substantially changed'. When it comes to the most extreme forms of weather, like typhoons and hurricanes, 'the theoretical maximum intensity is expected to increase with temperature,' but 'climate models give no consistent indication whether tropical storms will increase or decrease in frequency or intensity as climate changes' (*ibid.*: xxv). In any case, extreme weather involves pronounced social risks:

> A single [weather-related] disaster can completely negate any real economic growth for several years in a poor country, by diverting scarce domestic resources into relief and recovery programs and by increasing the burden of international debt. When severe typhoons strike islands in the southwest Pacific, it is not unusual for the entire annual agricultural output to be lost.
>
> (Mitchell and Ericksen 1992: 146)

Objections to the greenhouse theory

Are the risks associated with the enhanced greenhouse effect real? Basically, two doubts exist. First, the reality of an increase in global mean temperature precipitated by concentrations of heat-trapping gases is questioned. Second, even if warming takes place and causes climate change, the environmental effects may be manageable, even benign in some places. I shall examine these issues with a view to identifying major scientific controversies in the climate debate.

The debate got its impetus from the observation of a trend towards higher mean temperature, running parallel to growing emissions of greenhouse gases. But is such a trend really discernible in meteorological data? According to critics of the greenhouse theory, it does not show clearly.

> The data are ambiguous, to say the least. ... the greatest temperature increase occurred *before* the major rise in GHG [greenhouse gas] concentration. It was followed by a quarter-century decrease between 1940 and 1965, when concern arose about an approaching ice age! Following a sharp increase during 1975 to 1980, there has been no clear upward trend ... during the 1980s, despite some very warm individual years and record GHG increases.
>
> (Singer *et al.* 1993: 349)

Two additional problems with meteorological data have been noted. Global temperature measurements were commenced in 1860, during an especially cold period, and had the baseline been set at a much warmer period – the 1940s or the 1780s – the rate of heating would have been less pronounced (Singer 1992: 5). Furthermore, the recent rise of mean temperature may be inflated by the so-called urban heat island effect:

> Most temperature devices are located near large cities, where readings are distorted by concrete, roads, and other incidents of concentrated human habitation. When data taken at some remove from cities are analyzed, the apparent increase is not detectable. Ocean temperatures have not changed from 1850 to 1987. TIROS-N weather satellites, measuring global temperatures from the vantage point of space, have detected no evidence of a long-term global warming or cooling trend.
>
> (Stone 1993: 20)

If, despite these objections, a historic trend towards higher temperatures is granted, the next issue is its explanation. While the greenhouse theory traces it to emissions of heat-trapping gases, the IPCC acknowledges that an alternative explanation exists.

> The size of this warming [an increase of global mean surface air temperature by 0.3°C to 0.6°C over the last 100 years] is broadly consistent with predictions of climate models, but it is also of the same magnitude as natural climate variability. Thus the observed increase could be largely due to this natural variability; alternatively this variability and other human factors could have offset a still larger human-induced greenhouse warming.
>
> (IPCC 1990: xii)

It is known that global mean temperatures fluctuate. Among the many factors that cause such fluctuation are solar irradiance change and the cooling effects of stratospheric and tropospheric aerosols. If the observed increase of 0.3–0.6°C lies within the range of natural variability, it may be a matter of chance. This is the crux of the alternative explanation hinted at in the passage quoted above. It will be called the ***natural variability hypothesis***.

The IPCC deflates this hypothesis by suggesting that natural variability may have suppressed the average temperature by inducing cooler weather. If so, the anthropogenic greenhouse effect is greater than observed, part of it being masked by natural variability. Proponents of the natural variability hypothesis counter that such a lopsided pattern of temperature fluctuation is highly unlikely.

> [A] high proportion of simulations from the 'natural variability' theory give fluctuations equally spaced about a mean value.... the probability of 'natural variability' supplying a constant negative bias on all mean temperatures over a century or more of the 0.5°C or so required to justify the greenhouse theory is extremely low. We are therefore justified in concluding that the greenhouse effect has a very small chance of being responsible for the major part of global temperature change.
>
> (Gray 1992: 6)

But the extraordinary contingency of 'all the elements of variability maintain[ing] a constant negative value for periods of a century or more' (*ibid.*: 8) is no more extraordinary than their maintaining a constant positive value for such periods, which they would have to have done in order to produce the observed temperature increase of 0.3–0.6°C. Historical data neither confirm nor undermine the natural variability hypothesis.[4]

The greenhouse explanation derives, to recall, from a confrontation between meteorological data and *post hoc* predictions based on GCMs. I have already mentioned some shortcomings of GCMs, and critics of the greenhouse theory seize on these to reinforce their criticism. One objection, raised by Robert C. Balling (1992), is that the models do not yield accurate estimates of past temperature trends. Balling argues that the global mean temperature increase of the previous 100 years falls around 50% below GCM forecasts based on retrospective data on greenhouse gas emissions. This deviation bespeaks faults in the models. Some analysts go further, denying the scientific credibility of the GCMs altogether, on the ground that the attempt to model physical processes in the atmosphere is futile. The geologist Loren C. Cox (1991: 5) writes:

> The atmosphere is made up of many powerful forces which are interactive within the air mass, and these, in turn, variably interact with land and water masses over which the atmospheric column is moving. Temperature, water vapour, pressure changes, differential wind velocities at different altitudes and friction, all must be measured, and models must be attempted to show likely, near term future outcomes before physical movement rearranges many of the parameters. The task of such forecasting would be similar to an attempt to model, and predict with precision, the movements in one day of 100 stocks on the New York Stock Exchange – including their closing price. Such a predictive model is impossible to construct because of the wide range of variables that will affect the price movements of stock shares.

Besides branding existing models as untrustworthy, Cox implies that refinements in modelling can only marginally improve their predictive power. Atmospheric physics is simply too complex for this effort to succeed. When proponents of the theory claim that 'models are better than handwaving', critics retort: 'how much better?' (Singer *et al.* 1993: 351). If they are right, not only is the greenhouse explanation of past climate change undermined, but the greenhouse *prognosis* folds up as well. Extrapolating current levels of greenhouse gas emissions, the IPCC predicts an unprecedented rise in global mean temperature, but this is mere guesswork if the GCMs convey an overly simplistic representation of the causal mechanisms that link emission levels to climate developments.[5]

The final element of the greenhouse theory is the prediction that climate change will disrupt welfare in many parts of the world. Adverse social effects are believed to ensue from sea level rise, disturbances of ecological systems and a more turbulent climate. Critics of the theory first deflate the risk that land masses will be flooded if rising temperatures in high latitudes melt the Antarctic icecap. Christopher Stone (1993: 22) claims the better betting to be 'that global warming would *thicken* the ice caps and tend to lower sea levels'. The reason is that: 'a globally warmed environment would experience more evaporation and precipitation. Part of this hydrologic process involves removing water from the sea and depositing it, stowed as snow, on the polar ice packs – thereby thickening them' (*ibid.*: 23).

The critics also doubt that global warming will disturb ecological systems and, in particular, jeopardize agricultural production. Parts of the earth will rather benefit from a warmer climate. More arable land may become available in the northern regions of Russia, and there is a potential for more precipitation in the semi-arid tropics (e.g. India) because monsoonal activity will increase (Morrisette 1991: 157). Moreover, carbon dioxide has a fertilization effect that will augment agricultural crops, thereby offsetting negative effects of heating induced by greenhouse gases (Nordhaus 1991a: 45). Finally, the threat of more turbulent weather is questioned:

> It is the extreme climate events that cause the great ecological and economic problems: crippling winters, persistent droughts, extreme heat spells, killer hurricanes, etc. There is no indication from modelling or actual experience that such extreme events would become more frequent if GHW [greenhouse warming] becomes appreciable. The exception might be tropical cyclones, which ... would be more frequent but weaker, would cool vast areas of the ocean surface, and increase annual rainfall. In summary, climate models predict that global precipitation should increase by 10 to 15%, and polar temperatures should warm the most, thus reducing the driving force for severe winter-weather conditions.
>
> (Singer *et al.* 1993: 352)

In sum, all elements of the greenhouse theory are surrounded by scientific controversy. Critics reject its explanation of the past and, *a fortiori*, have no trust in its warnings about the future. Whether or not their arguments succeed in rebutting the greenhouse prognosis, they leave little doubt that current knowledge of climate change is imperfect in many respects. In what follows, I shall scrutinize the scientific debate with a view to devising a more precise diagnosis. Before that, a brief overview of the main scientific controversies will be useful.

(i) Do meteorological data show an increase in global mean temperature over the past 100 years?
(ii) Does the greenhouse theory explain the past temperature pattern better than the natural variability theory?
(iii) Are GCMs adequate representations of physical processes in the atmosphere?
(iv) Will climate change have disruptive social effects?

The majority argument

Contentious issues notwithstanding, the greenhouse theory comes close to passing for conventional scientific wisdom. It is a natural point of departure when it comes to taking stock of what we know about climate change. Does the theory merit this privileged epistemic position? One reason for thinking so is that the greenhouse theory is the *majority* view. William R. Cline (1992: 3) argues:

> The public policy debate on the greenhouse problem has often given the impression that scientists are hopelessly divided. Instead, the IPCC scientific report provided an important occasion to reveal a wide consensus among a substantial majority of scientists, especially those directly trained in climate sciences. The report, called for by the United Nations, assembled an international group of 170 scientists, as well as another 200 reviewers. It would be fair to say that the scientific majority accepting the report's findings is at least as large as the two-thirds threshold required for proposing constitutional amendments in the United States.

There is, however, an important difference between political decision making based on democratic procedures and the choice of whether or not to put one's trust in a scientific theory. Within a democracy, every person beyond a certain age has an equal say in the selection of representatives to the legislative assembly, and every member of that assembly has an equal say in its decisions. Such egalitarianism is out of place in the world of science. Descartes (1628/1967: 6) remarks: 'if it is a question of difficulty that is in dispute, it is more likely that the truth would have been discovered by few than by many'. This is not a plea for minority rule in scientific matters, but a reminder that here only the opinions of reputable experts count. Cline's argument must be revised accordingly, and here is a revised version of it, to be called the *majority argument*:

(i) If a substantial majority of scientific experts concur in an opinion within the field of their expertise, this opinion is more credible than alternative points of view.
(ii) A substantial majority of scientific experts on the climate accept the greenhouse theory.
(iii) The greenhouse theory is credible.

We may start by examining postulate (ii). It is, on the face of it, a noncontroversial premise. A large number of scientists in fields that are relevant to the study of climate change have taken part in the work of the IPCC, and only a small minority of reputable experts launch serious objections. It has been charged, to be sure, that the IPCC is politically biased and its views unrepresentative of mainstream professional expertise, but the sheer number of scientists involved (170) gives reason to doubt this, and an extended analysis of the recruitment process finds nothing to support the charge (Lunde 1991). Yet, doubts are raised about the extent to which scientists who served as authors or referees of the IPCC reports actually endorse the greenhouse theory. An inquiry among contributors from the United States, carried out in 1991, shows that 'Only 15 percent believed that the current GCMs accurately portrayed the atmosphere–ocean system, and less that 10 percent thought that current GCMs had been adequately validated by the climate record' (Singer 1992: 9). In view of the fact that only 46 of the 126 questionnaires sent out were returned, this result does not undermine the claim of the IPCC to represent a majority view. It indicates, however, that the guarded language of some of its general conclusions may patch over substantial disagreement about specific premises. But without more detailed knowledge of how dissenters think, little can be made of their dissent when it comes to epistemic diagnosis. Postulate (ii) may therefore be granted.

What about (i), which is the basic premise of the majority argument, to be called hereafter the *majority postulate*? If a substantial majority of scientific experts concur in an opinion within their field of expertise, is this opinion more credible than alternative points of view? No, says the leading opponent of the greenhouse theory, denying that credibility is correlated with numerical support.

> An agreement among the majority of scientists does not mean that the majority is correct. It only means that the majority of scientists agree on something. The minority may well be correct; it could be a minority

> arguing for a very large increase in temperature (a catastrophic increase, if you like) or a minority arguing that very little will happen.
>
> (Singer 1989: 3)

The weak point in this argument is the dismissal of majority agreement as a mere fact of the majority agreeing on something. First, there is no more reliable source of knowledge on a given phenomenon than leading academic authorities in the relevant field. Those who think otherwise are apt to reject the whole body of modern science, and, professing such radical scepticism, they are apt to see the approach to epistemological problems taken in this book as fundamentally flawed. Second, if opinions diverge among equally competent and trustworthy scientists, the view that draws the assent of most is most likely to be correct, especially if it has a substantial majority on its side. Scientists are not, of course, infallible, and history tells many tales of misplaced trust in professed scientific experts, but scientific opinion is, *faute de mieux*, our best guide.

This is not to deny that psychological, economic and political factors sometimes influence the prominence of a position within the scientific community, rendering it more or less popular than it ought to be on purely intellectual grounds.[6] But this is unlikely to happen with highly publicized positions that have prompted sustained critical assessment. An opinion that has a constant attraction for most scientific experts even under such conditions merits more confidence than other opinions because the best explanation of its special standing is that it has a balance of reasons on its side. The minority may well be correct, but if one must place one's bets somewhere, they are better placed with a large majority. Moreover, deferring to the majority in scientific matters is not tantamount to ignoring minority opinions. It would, for example, be irrational to brush aside a small number of reputable experts who insist that a certain activity is very dangerous even if most experts pronounce it safe, and equally irrational to overlook exceptional optimists when the prevailing view is pessimistic. But it is not irrational to have greatest confidence in the theory that commands widest assent.

This will not be the last word on the validity of the majority postulate. For the moment, however, let us assume that the postulate is correct. If so, the greenhouse theory is credible. What does this imply as far as our knowledge of climate change is concerned? First, the risk that continued emissions of anthropogenic greenhouse gases at present rates will alter the world's climate is greater than the likelihood that they will not. This is an *ordinal* probability estimate: one outcome is deemed more probable than another. In general, ordinal comparisons of the likelihoods of various states of affairs can be made if these states can be ranked in terms of how likely they are. As far as the climate is concerned, I believe it more likely to change than not to change owing to unabated greenhouse gas emissions, because most scientists see things this way.

A further question is whether anything can be said about the difference in likelihood between the two outcomes. Does knowledge diagnosis permit an estimate, if only in non-numerical terms, of how more likely the climate is to change than not to change? The small number of vocal critics of the greenhouse theory might indicate that change is a *much* more likely outcome. Such an inference would, however, be fraught with arbitrariness, and I shall set no store by it.

The sceptical argument

Is the majority postulate valid and a proper basis of knowledge diagnosis? According to one view, the existence of a minority of reputable experts in a given field undermines the credibility of all scientific opinions. Thus, scientific opposition to the greenhouse theory casts doubt on every prediction – frightening or not – that scientists have come up with. Some believe, indeed, that 'the level of all ecological knowledge is primitive, and the climate is no exception' (Easterbrook 1992: 21).

General premises on which to base such a view can be found in an article by Dagfinn Føllesdal (1979). Its subject is the risks associated with genetic research. Føllesdal asks: should genetic research which most scientists believe to be very beneficial but some consider risky – notably recombinant DNA research – be permitted? In one respect, this problem is an exact opposite of the problem of climate change. While the majority of qualified experts see a high risk of unprecedented warming in emissions of greenhouse gases, the warning about DNA research has less support – and, hence, less initial credibility – than predictions to the effect that such research is safe. In another respect, however, the problems of DNA research and climate change are alike. Theories that are used to predict the outcomes of genetic experiments are as inchoate as the greenhouse theory. Føllesdal (*ibid.*: 405) asserts that we 'know at present little concerning the function of the various genes, and how they interact with one another,' and that cannot exclude the possibility 'that small changes may bring forth big and unknown effects'. In general, where 'we have several competing theories, which give different predictions, all these theories should be regarded with suspicion and we should be prepared for a risk that is higher than what is predicted by any of the theories' (*ibid.*: 405–6).

Føllesdal takes disagreement among experts to indicate that a given phenomenon lies beyond our full grasp. When divergent predictions are all put forth with some credence, the credibility of every prediction is undermined. This will be called the *sceptical argument*. Basically, it says:

(i) Theoretical controversy casts doubt on all theories alike.
(ii) The chance that things will go wrong may be greater than any theory predicts.

Proposition (ii) follows from proposition (i), which also entails:

(iii) The chance that things will go well may be greater than any theory predicts.

If, that is, theoretical controversy undermines our confidence in scientific predictions, it will have to cut both ways, casting doubt on alarmist as well as sanguine predictions. We may err on the side of both pessimism and optimism.

The fact that a minority of scientists reject the greenhouse theory is, according to the sceptical argument, a sign that science has yet to disclose the true nature of climate change. All existing theories are controversial, and every probability estimate, even at the ordinal level, should be hedged with reservations. No one can be trusted as long as dissent is voiced from the ranks of scientific experts, because the best explanation of conflicting opinions about a phenomenon is that nobody fully comprehends climate processes. For all we know, an unprecedented warming may be under way; it might even be worse than proponents of the greenhouse theory predict. But it is also conceivable that emissions of greenhouse gases affect the global mean temperature very little. As there is no basis for estimating the probability of either possibility, we cannot say whether the greenhouse prognosis is more likely to be true or false. The climate is, in brief, a big unknown.

The majority postulate and the sceptical argument are grounded in diametrically opposed approaches to knowledge diagnosis. The former takes the existence of a substantial majority to all but overrule the authority of the minority; the latter sees the existence of a minority as a blow to our confidence in any scientific opinion. The previous discussion shows that something can be said for both views, and that neither stands out as *the* proper basis for diagnosing current knowledge of climate change. We have, in other words, come to a standstill. I shall try to bypass it in the next section.

The piecemeal approach

Both arguments of the previous section judge the greenhouse theory in one lot. It may seem that a more proper approach is proceeding piecemeal, not assessing a theory wholesale but scrutinizing its various components individually to arrive at an overall assessment.[7] Thus, Andrew R. Solow (1991: 25) writes that, 'Based on a balanced reading of the scientific literature, it is virtually certain that global warming will occur in response to ongoing changes in atmospheric composition'. The problem is only, as so often when someone claims to have seen through a controversy and come to a balanced conclusion, that the premises remain obscure. I shall try to be more precise.

A convenient point of departure is the four scientific controversies that were identified at the end of the section 'Objections to the greenhouse theory'. Each will be examined again with a view to determining the credibility of the greenhouse theory. Central to such an examination are, first, the substantive arguments put forth by proponents and opponents of the theory, and, second, the previous reflections on the epistemological status of the substantive arguments. I shall call them *first-* and *second-order* considerations, respectively. The latter will be brought into the assessment of the greenhouse theory together with a discussion of substantive matters, and their role in knowledge diagnosis may vary from one issue to another. It is conceivable, in other words, that the majority postulate bolsters the credibility of the greenhouse theory on one controversy, while the sceptical argument undermines it with respect to another.

The first controversy

The first controversy has to do with the reading of meteorological data. Do they or do they not show an increase in the global mean temperature over the present century? The importance of this issue is obvious. If no temperature increase has so far resulted from anthropogenic emissions of greenhouse gases, the worry about an enhanced greenhouse effect seems unfounded. Existing data are ambiguous. The IPCC claims to have identified an increase of 0.3–0.6°C over the last 100 years, but there are some conspicuous deviations from a smooth curve, notably the cooling that took place between 1940 and 1965. Critics of the greenhouse theory argue, moreover, that the historical record is distorted by the facts that the cold 1860s serve as a historical baseline, and temperature devices are mostly located near large cities housing many activities which tend to heat the air.

The latter problem may be serious, but must still be put to one side. Faulty or dubious data are a common phenomenon, but so is the necessity of basing empirical analysis on such data for lack of better ones. The former problem – the erratic nature of the trend towards a higher global mean temperature – can also be set aside, but not because it defies solution. It rather takes us right into the second controversy.

The second and third controversies

There is no disagreement on the existence of an erratic trend, only on its explanation. The real dispute is whether the trend can best be explained by the greenhouse theory or by the natural variability hypothesis.

Both explanations fit the facts, if only in conjunction with auxiliary hypotheses. The greenhouse theory may need help to account for pronounced fluctuations around a rising temperature curve, the natural variability hypothesis to accommodate the fact that there is a rising curve amid fluctuations. How are advocates of either explanation to account for random factors that may have affected global mean temperature over an extended period, constantly suppressing it or gradually pushing it upwards?

As far as the greenhouse theory is concerned, a plausible auxiliary hypothesis is that sulphate aerosol compounds exert a cooling effect by shielding the earth's surface from sunlight. The *1992 IPCC Supplement* stresses the role of aerosols in making parts of the globe significantly colder than they would have been without this counteraction of the enhanced greenhouse effect. Thus, the cooling that was recorded between 1940 and 1965 may be an extreme manifestation of the constant tendency of aerosols to flatten the temperature curve. Recent research lends credence to this hypothesis (Charlson *et al.* 1992; Kiehl and Briegleb 1993; Charlson and Wigley 1994). The direct cooling effect of sulphate particles is comparable in magnitude to the heating caused by carbon dioxide, at least over industrial regions. Assuming that GCMs correctly estimate the contribution of carbon dioxide to global warming, the evidence indicates that 'from 1880 to 1970, aerosol cooling may more or less have canceled out the enhanced greenhouse effect in the Northern Hemisphere' (Charlson and Wigley 1994: 34). The counteracting properties of aerosols may therefore explain why the warming trend of the early decades of this century ceased around 1940, to resume only in the 1970s.

When it comes to the hypothesis of natural variability, what seems to be needed is an auxiliary hypothesis explaining why the global mean temperature has increased amid fluctuations over the last 100 years. But is it really needed? Advocates of the natural variability hypothesis may not think so. They may refuse to see a positive bias calling for special explanation in a temperature increase of only 0.3–0.6°C. The historical record does not deviate significantly from a pattern of equally distributed fluctuations around a consistent middle value.

This argument proceeds too fast, however. The natural variability hypothesis needs the reinforcement of an auxiliary hypothesis if there is something to the claim that increasing concentrations of sulphate aerosols have in fact caused a constant cooling effect in the northern hemisphere. Granted this negative factor and furthermore assuming – as advocates of the natural variability hypothesis do – that no enhanced greenhouse effect has made itself felt, the observed temperature increase of 0.3–0.6°C is anomalous. Thus, advocates of the natural variability hypothesis must either (i) deny the existence of a cooling effect caused by aerosols, or (ii) establish that the observed global warming is no more likely to stem from an enhanced greenhouse effect than to reflect random factors counteracting the cooling effect of aerosols.

As far as I know, (i) has not been attempted, but (ii) will be accomplished if objections to the use of GCMs in climate research are valid. If, that is, the enhanced greenhouse effect is posited on inadequate models of the causal relationship between greenhouse gas emissions and climate change, as critics of the greenhouse theory say, there is no more reason to believe that such an effect exists than to assume that random factors have produced the temperature increase. It is, indeed, more likely to have been brought about by random factors than a momentous but purely conjectural atmospheric mechanism linking greenhouse gas emissions to climatic change. If GCMs do nothing to validate the existence of an enhanced greenhouse effect, the greenhouse theory has the uncomfortable appearance of a *deus ex machina*. If, by contrast, GCMs provide an adequate (if not accurate) representation of the atmosphere and so give reason to believe that the enhanced greenhouse effect exists, then the greenhouse theory explains the pattern of the global mean temperature over the past century better than the natural variability hypothesis, offering a definite clue as to why warming has taken place despite the cooling effect of aerosols.

The argument of the previous paragraph turns on the third controversy: are GCMs adequate representations of physical mechanisms in the atmosphere? If they are, the natural variability hypothesis needs an auxiliary hypothesis to explain the rise of the global mean temperature over the last 100 years; if they are not, the natural variability hypothesis is, despite this rise, more credible than the greenhouse theory. Expert opinions are diametrically opposed on this score. Critics of the greenhouse theory find the models wholly inadequate and beyond repair; proponents of the theory acknowledge their shortcomings but vouch for their overall validity. The former believe, as we have seen, that the models are little better than hand-waving; the latter hold that 'the improvements made in model structure and sophistication in the last few decades and the various tests that have been performed using these models have produced a growing sense that, despite their remaining flaws, the models have reached a useful degree of realism' (Firor 1990: 84).

Those who are not versed in atmospheric physics can add nothing to this discussion. We may, however, turn to second-order considerations that bear on the epistemological status of competing substantive arguments. The fact that the largest number of scientific experts vouch for the GCMs indicates that they are more likely to be realistic than fundamentally flawed – or so, at any rate, the majority postulate suggests. But the lack of scientific agreement is a sign that no one understands the relevant atmospheric processes and – if we are to believe the sceptical argument – that no effort at modelling these processes has any credibility. Which second-order consideration has more to be said for it in this case? One reason to reject the sceptical argument is that scientific modelling is normally fraught with controversy. If disagreement on the realism of models is taken to imply total ignorance of the phenomena meant to be portrayed, the lion's share of current scientific knowledge goes by the board. The general implications of scepticism on this score are calamitous, if not absurd. This argument may meet with the objection that controversies over models vary in seriousness. Some run deeper than others, and only the most profound give reason to infer from controversy total ignorance about the substantive matter at hand. This opens the door to selective application of the sceptical argument that may largely deflect its ruinous implications for scientific knowledge. It is arguable, in particular, that the controversy over GCMs is especially serious, and that it is not the mere fact of disagreement surrounding GCMs which betokens their inadequacy, but the extent to which scientific experts disagree. The sceptical argument applies to this case because it is a special case.

Can we draw a clear line between more or less serious cases of scientific controversy? The distinction is most elusive. Thus, the distance between endorsing and rejecting GCMs is not obviously wider or narrower than, say, the distance between geocentric and heliocentric models of the universe in the sixteenth century or the distance between psychoanalytic and rationalistic models of human behaviour today. Precise judgements can hardly be made on this score, and the room for arbitrary ones is too great to permit discriminate application of the sceptical argument.

This conclusion renders the majority postulate directly relevant to the third controversy. If, amid controversy, we balk at the conclusion that every opinion is equally untrustworthy, it is natural and rational to put greatest trust in the opinion of a large majority of reputable experts. Second-order considerations suggest, in other words, that GCMs are more likely to capture atmospheric processes in an adequate manner than to be fundamentally flawed. This implies, in turn, that the greenhouse theory has more credibility than the natural variability hypothesis. The former explains both why the global mean temperature has increased by 0.3–0.6°C over the last 100 years, and – in conjunction with the auxiliary hypothesis about aerosol-induced cooling – why the increase has neither been smooth nor uniform over the globe. By contrast, the natural variability hypothesis cannot explain why the temperature curve points upwards despite the occurrence of a powerful negative component among the sources of fluctuation.

The foregoing argument indicates that the likelihood of global warming and climate change exceeds the likelihood of the world's climate staying stable over the next century. This is not to say, of course, that warming will inevitably result if nothing is done to curb emissions of greenhouse gases, but from what we know today, pessimism is the proper attitude on this score. The greenhouse prognosis is more likely to be true and climate change will likely result from greenhouse gas emissions.

The fourth controversy

The final component of the greenhouse theory is the prediction that large parts of the world will experience adverse social effects in the wake of climate change. Life conditions will deteriorate as a result of rising sea levels, ecological disruptions and extreme weather. This controversy turns on the validity of this prediction. If invalid, the conclusion of the previous paragraph, which takes climate change to be the more likely outcome of unabated greenhouse gas emissions, gives no reason for alarm.

An obvious problem in ascertaining where the truth lies is that the element of guesswork looms large in every prediction of how people will fare in a warmer climate. The nexus of causes is extremely complex, involving an array of natural and social mechanisms. Forecasts cannot be derived from an explicit model of relevant factors, not even a primitive one, but owe a lot to imagination. It is clear, however, that the greenhouse theory has not been rebutted on this point. Doubt has been raised about it – enough to conclude that precise predictions of environmental and social effects cannot be made – but the prospects of rising sea levels flooding coastlands, ecological changes undermining agricultural production or weather conditions becoming very turbulent are not pulled out of thin air. One cannot, for example, lightly take on assessments to the effect that 'sea level rise induced by climate change may force the relocation of up to 80 million people [in Bangladesh], reduce rice-producing land by up to 2.6 million hectares and projected rice output in 2010 by 8–15 per cent' (Hayes 1993: 129).

The next question is whether or not the risk that climate change will disrupt welfare is greater than the likelihood of bearable social effects. This brings by now familiar epistemological considerations to the fore. On the one hand, a majority of climate scientists, whose view finds expression in the report of the IPCC, are pessimistic. On the other hand, the conjectural nature of research in this field, avowed even by analysts who generally endorse the greenhouse theory, might suggest that no scientific forecast merits much confidence. Once more, in other words, we are up against the conflicting implications of the majority postulate and the sceptical argument. But this is not a case of distinct theories offering coherent but divergent understandings of the same phenomenon. What we have are inchoate and largely discrepant analyses of different aspects of the relationship between the climate and human living conditions. Thus, we should not set too much store by the sheer number of scientists believing or doubting that climate change will significantly worsen these conditions, as their beliefs do not reflect the convergence of opinion around a well defined scientific position. Nor, however, should disagreement with respect to the relationship between climate change and social welfare undermine our confidence in scientific forecasts that derive from observations of the vulnerability of certain regions to extreme weather or climate-induced alterations of the sea level or ecological parameters of agriculture.

In the present state of knowledge, I can think of only one argument that might justify writing off the adverse welfare implications of climate change. It might be done on grounds of the human potential for adapting to environmental disruption. Robert J. Samuelson (1992: 32) asserts, for example, that 'the warming would occur over decades. Peoples and businesses could adjust. To take one example: farmers could shift to more heat-resistant seeds'. This argument rests crucially on what we may call the *presumption of progress* – the idea that, given time, human beings have a considerable propensity for developing means to cope with emerging threats to their welfare. They will, more specifically, be better situated to cope with climate change in the future than anyone can imagine today.

There is much to be said for the general validity of the presumption of progress. (I shall return to it in chapter 5.) Human beings display a pronounced propensity for coping with environmental adversities, owing largely to technological innovation. True, innovation poses a specific problem of knowledge. Its pace and direction some years hence is quite unpredictable. We know nothing, for example, about the ingenuity or despair of future people in mending ecological damage. But innovation takes place all the time and history abounds with successful efforts to overcome adversities that seemed insurmountable until some novel means unexpectedly emerged. The presumption of progress stands for more than mere hope. People will probably be better situated to cope with climate change in the future than they are today. Still, the general validity of the presumption notwithstanding, serious reservations may be had about its relevance in the present context. First, the greenhouse prognosis heralds swift changes in the world's climate. The rate of warming is predicted to reach dangerous levels within five to eight decades if nothing is done to slow down the concentration of greenhouse gases in the atmosphere. This leaves little time for epoch-making innovations. Second, the effects of climate change may touch on things that have proven very resistant to change throughout history. The roles of freshwater and arable land in sustaining human life come close to being constants of the human condition. One may acknowledge historical progress and still deny that rising sea levels or the fast emergence of a dry and barren environment will make little difference to people's future welfare. This is not to deny that the historical record is mixed as regards the propensity of climate change to outdistance human adaptability.

> Some historians claim to see the effect of climate in events as great as the decline of the Late Bronze Age civilization of Mycenae three thousand years ago and the much more recent transatlantic movement of European populations; other scholars point out that other trends were also under way that could equally well have caused the cultural or economic shifts.
> (Firor 1990: 67)

But this only indicates that human adaptation to environmental transformation is neither inevitable nor a foregone conclusion, and environmental adversities of the kind envisaged by the greenhouse theory will put those affected to a very hard test. In view of this, the warning that climate change will disrupt welfare has a better chance of being true than optimistic reassurances to the contrary. Should sea levels rise, ecological disruptions or turbulent weather occur, the social effects will most likely be adverse, and current knowledge gives no reason to assign a lower probability to the occurrence than the nonoccurrence of these environmental effects of climate change. Nor, however, are we warranted in going beyond an ordinal estimate of relative likelihood.

Here ends the piecemeal approach to diagnosing knowledge about the risk that unabated emissions of greenhouse gases will cause climate change which undermines living conditions in parts of the world. The conclusion is twofold: first, unabated emissions of greenhouse gases are most likely to push up the global mean temperature and trigger climate change; second, the risk that climate change will have adverse social effects in many areas exceeds the chance of successful human adaptation to the environmental consequences of this change.

The risk of reducing greenhouse gas emissions

The previous discussion proves the reality of one horn of what might be an environmental dilemma: if atmospheric emissions of greenhouse gases continue unabated, the result may be environmental damage that makes some people worse off than they would be if emissions are reduced. Is there another horn? Is there, in other words, a real risk associated with reductions of greenhouse gas emissions?

To lower atmospheric concentrations of greenhouse gases, carbon dioxide emissions must be curbed, which in turn requires scaling down the use of fossil fuels. How will efforts to this effect influence the standard of living? In particular, will cuts in energy consumption quell economic growth? The rate of growth depends on an array of factors that combine to create a complex causal nexus, and there is ample reason to treat predictions in this field with caution. Thus, William Colglazier (1991: 65) asserts that the 'debate over the cost of cutting greenhouse gas emissions is reminiscent of the debate a decade ago over the cost of reducing oil imports, and it is worth remembering that some of the policy responses then (e.g., price controls and allocations) exacerbated the problem'. In this respect, however, the economic debate is no different from the debate about atmospheric effects of greenhouse gas emissions. Both reveal a future that is subject to considerable ignorance. Moreover, revelations of the cost of adjusting economic activity to stabilize greenhouse gas concentrations vary with variations in the method of analysis. I shall present three approaches and an attempted synthesis of these.

In the economic profession, *top-down* analysis is the predominant approach. It forecasts future economic developments by means of macroeconomic models which start from assumptions about the contribution of energy to economic output and proceeds to estimate the cost of specific carbon dioxide reductions by stipulating the economic effect of corresponding reductions of energy use. Economic theory

> states that the 'elasticity' of output with respect to a productive 'factor' is the factor's share in production value. Energy commands a share of 6 to 8 per cent of GDP. Theory thus suggests that the elasticity of output with respect to energy is about 0.08. That is, a 10 per cent reduction in energy should cause a 0.8 per cent reduction in GDP.
>
> (Cline 1993: 11)

It may reasonably be expected that such analysis yields pessimistic conclusions about the costs of curbing carbon dioxide. Implementation costs may be high and regulations hurt productivity and thereby depress economic output. As it turns out, however, this hypothesis finds no support in forecasts of the effect of moderate carbon dioxide reductions. According to William D. Nordhaus (1991b: 61), a 10% lowering of greenhouse gas emissions can be obtained at a cost of less than $10 per ton of carbon dioxide equivalent, but large-scale reductions are another matter.

> After 20 percent reduction ... the curve [of both marginal and total cost] rises as more costly measures are required. A 50 percent reduction in GHG emissions is estimated to cost almost $200 billion per year in today's economy, or around 1 percent of world output. This estimate is understated to the extent that the implementing policies are inefficient or that they are undertaken in a crash program.
>
> (*Ibid.*: 63)

On this basis, Nordhaus (1990: 20) sees a major economic downturn as a result of implementing the IPCC proposal for abatement measures aimed at stabilizing greenhouse gas concentrations. The proposal implies that carbon dioxide emissions must be more than halved before the middle of the next century, and this would 'plunge the world into depression' (*ibid.*: 21).

The top-down approach to estimating abatement costs is criticized on the ground that it presupposes 'a behaviourally, institutionally and technologically fixed world' (Wilson and Swisher 1993: 250). The alternative is *bottom-up* analysis – a disaggregate approach that includes conjectures of how far technological and institutional innovation will lower the future costs of carbon dioxide reductions.

The most optimistic studies adopting this approach 'conclude that much can be done to mitigate global warming at little or no cost to society' (*ibid.*: 249). Others, however, are less sanguine. Cline (1993: 14–15) argues that carbon dioxide emissions could be cut by around 20% at zero cost 'as an initial tranche of abatement', but cautions that this 'move to technical efficiency' will be a 'one-time shift that is relatively minor when placed in the perspective of the long-term trend in emissions'.

Both top-down and bottom-up analysis may be supplemented with estimates of the contribution of forestry measures to lowering the carbon dioxide content of the air. Because growing trees absorb carbon, afforestation – or, more modestly, reduced deforestation – is a means of counteracting the enhanced greenhouse effect. Its effect is, however, temporary. Carbon dioxide is not absorbed by trees that have ceased to grow and is emitted by dying ones. Yet afforestation 'might provide an important window of policy response over some 30 to 40 years and thereby buy time for technological change in noncarbon energy sources to take place' (Cline 1992: 217). Incorporating this assumption in a major study that otherwise combines top-down and bottom-up analysis, Cline (*ibid.*: 232) predicts that 'carbon reductions of some 50 percent to 60 percent ... may be purchased at some 2 percent of world GNP or less at the middle of the 21st century (and considerably more cheaply at the beginning of the century because of start-up gains from afforestation)'.

Current research indicates, accordingly, that eliminating up to one-fifth of existing carbon dioxide emissions is unlikely to hurt the world economy very much, but economic adversity will most likely result if emissions are more than halved. This is to say that a full implementation of the cutback proposed by the IPCC (1990) renders the risk of economic decline higher than the chance of continued growth. Although economic analyses, irrespective of method, are exploratory and conjectural, broad agreement exists on this ordinal probability estimate.

This proves the reality of the second horn of the environmental dilemma involved in choosing climate policy. There are true risks associated with both options before us: unabated greenhouse gas emissions as well as drastic reductions. The next question is what to do. Now, that question cannot be answered before normative standards for the resolution of environmental dilemmas have been worked out. After an effort to this effect has been made in chapters 2 through to 5, I shall return briefly to the problem of climate policy in chapter 6.

General lessons

Most environmental dilemmas involve risk. We have to choose whether or not to engage in activities that may cause damage to the environment, and either alternative is seen as a threat to someone's standard of living by someone whose assessment on this score cannot be ignored. The first task is, then, to ascertain whether the risks are real or not. This chapter has shown how this should be done with respect to an environmental issue that currently attracts great attention.

If we stand back from the intricacies of climate change, some general lessons about risk analysis emerge from what was said above. The argument that underlies a particular assessment of risk should be taken apart, as it were, to determine how much trust one ought to put in its various components. Such investigations must take account of both first-order, substantive considerations pertaining to the validity of each part of a given argument, and second-order considerations that turn on the overall credibility of various arguments as judged by their standing among scientific experts. On the latter score, two guidelines were invoked: the idea that what a substantial majority of experts believes to be true, most likely is the truth, and the notion that disagreement among experts makes all opinions unworthy of trust. These guidelines were called, respectively, the majority argument and the sceptical argument. They come into the picture only if substantive considerations do not suffice to sort out a problem of knowledge. As they conflict, they open the door to different conclusions, and to determine what weight either argument ought to be accorded in particular cases, the following rule of thumb was suggested: the majority argument seems most germane if scientific disagreement consists in distinct theories offering coherent but divergent understandings of some phenomenon, the sceptical argument more to the point if opinions are generally inchoate.

Risk analysis is, as I said, the first task that confronts us when we stand before an apparent environmental dilemma. If such analysis leads to the conclusion that the dilemma is real, the next task is finding out how it should be resolved. Chapters 2 through to 5 work out normative standards that should guide our deliberations. I will briefly return to climate policy in chapter 6, which explores the implications of the standards in question for the dilemma that has been diagnosed above.

Notes

1 It has been argued that global warming may undermine the most important driving force behind the self-correcting process. 'The largest biological system on the planet is that of marine phytoplankton; it produces more biomass – 104 billions tons of carbon per year – than all terrestrial ecosystems combined, which generate 100 billion tons of carbon annually. Any reduction of photosynthetic activities in the phytoplankton could amplify global warming.... [I]t might provide less dimethyl sulfide, a gas which generates condensation nuclei for the formation of clouds' (Leggett 1992: 31). According to Stephen Schneider (1989, p. 103), experts have argued that positive feedback (which augments rather than restrains warming) could 'double the sensitivity of the climate system to initial injections of greenhouse gases,' but he adds that it is at present 'impossible to verify this frightening possibility'.

2 According to John Firor (1990: 56), the 'calculations required to capture these complexities involve simulating not only all the important processes at work in the air at one moment, but also the changes in these processes every few minutes. Such climate models require teams of scientists and computer programmers, working for years, for their design and construction, and millions of dollars' worth of computer time for theory testing and operation.'

3 It is 'the poorer nations with densely populated coastal areas that are most at risk. Some of the countries most vulnerable to sea-level rise include Bangladesh, Egypt, Pakistan, Indonesia, and Thailand, all comprised of large, poor populations. Indonesia, with 15% of the world's coastlines, is projected to lose 40% of its land surface should a one meter increase in sea level occur.... A one-meter increase could inundate 15% of Bangladesh and all of the Republic of Maldives, Kiribati, the Marshall Islands, Tokelau, Tuvalu, and the Torres Strait Islands' (Howarth and Monahan 1992: 10; cf. Mintzer 1992: 3).

4 I am grateful to Lars Fjell Hansson for helpful comments on this point. Another alternative to the greenhouse explanation of past temperature trends is invoked by Hugh W. Ellsaesser (1992: 2–3), who attributes a 'progressive, if not continuous, warming since the middle of the 18th century' to 'a recovery from a colder period, termed The Little Ice Age ... which began circa 1450'. Andrew R. Solow (1991, p. 21) concurs, arguing that the 'historic warming is consistent with a natural recovery from the Little Ice Age'. This explanation is glossed over by the IPCC, and, for want of more information, I shall do likewise.

5 Another challenge to the greenhouse prognosis may be mentioned in passing. Tor Ragnar Gerholm (1992) questions the credibility of assumptions to the effect that the total carbon dioxide equivalent of anthropogenic gases in the atmosphere may double during the next century. This presupposes, he says, that the global per capita consumption of energy will have become higher than today's European average by the year 2100, but, in view of economically recoverable resources of currently used fossil fuels, consumption cannot surge this high, and atmospheric concentrations of carbon dioxide will, accordingly, rise much less (*ibid.*: 6–8, 14–16). Now, the extent to which greenhouse gas emissions will increase in the next century if no drastic effort is made to curb them depends mainly on (i) how fast energy demand will rise in developing countries, (ii) how far it will be directed towards fossil fuels, and (iii) how far resources will be available to meet growing fuel demands. As energy analysts disagree on all scores, one should be wary of placing one's trust anywhere. It is a safe bet, however, that developing countries on the verge of industrialization, notably China, will soon contribute substantially to expanding global energy demand. Moreover, the economics of fossil fuel extraction place no fixed constraints on recoverable resources. Their size is an increasing function of their price and, hence, of the demand for them. Technological barriers to the exploitation of a particular resource at any given time are liable to recede through technological innovation when economic prospects become more favourable. It is, accordingly, neither impossible nor improbable that the global average of per capita energy consumption will match today's European average before the year 2100. Pending radical reorientations of industrial strategies in developed as well as developing countries, continued increase of fossil fuel use and growing emissions of carbon dioxide over the next century are a likely scenario.

6 Such mechanisms may not only give undue prominence to an opinion that already has broad backing, but also bolster minority views unreservedly: 'policymakers have occasionally indulged in self-serving efforts aided by news media, to exacerbate or amplify differences of opinion among scientific experts. There is a disquieting tendency extending from the congressional hearing room to television newsrooms to give equal time to the advocates of round earth science and spokespersons from the Flat Earth Society. While the minority view in greenhouse science must not be dismissed as Flat Earth fantasy, there is no question that it has gained far more access and influence in White House deliberations under [President] Bush than has the majority of scientists' (Hempel 1993: 232).

7 This approach was first suggested to me by Nils Roll-Hansen, who may not, however, agree with everything I say in the following.

2

The theory of interest and the priority principle

Introduction

There are many ways damage can be inflicted on the environment – forests can be razed, animal species exterminated, minerals depleted, and so on. Such damage does not always cause widespread concern. If, say, a mineral resource like chromium were rapidly being depleted but could be replaced with an adequate manufactured substitute, the prospect of being without chromium would hardly spread alarm and despondency. And if the ozone layer did not serve an important function in shielding human beings from solar radiation, the risk of its disintegration would probably be accompanied by indifference. In general, most people become concerned about damage to the environment only when they see it as a threat to the interests of human beings.

I shall not probe the justification of the anthropocentric attitude until the last chapter of this book. The present chapter argues only that environmental degradation is, indeed, a proper source of concern if it threatens human interests. It remains to be seen whether or not this is the most important reason for being concerned about it.

This chapter lends structure and justification to the idea that the environment should be conserved insofar as people have an interest in conserving it. I shall argue that human interests can be ranked in terms of normative force. What may be called *vital needs* – the physical prerequisites of survival and normal biological functioning – stand out by virtue of how much, objectively speaking, their fulfilment matters. From this argument, the following principle will be derived: it is imperative not to engage in activities whose impact on the natural environment deprives or risks depriving someone of the means of meeting vital needs, unless the risk is too small to be a reasonable source of concern. First of all, however, the method behind normative argument will be set out.

Method

We want to find out what we are justified in doing to resolve environmental dilemmas. Our aim is to arrive at normative standards that tell right from wrong. How can such standards be worked out? What is the proper method of normative thinking?

It is arguable that questions of method need not play a central role in moral philosophy. Brian Barry (1989: 258) writes:

> We all know how to engage in moral arguments, even if we would be flummoxed by being asked whether or not we subscribe to moral realism, objectivism, subjectivism, prescriptivism, or what have you. It is, moreover, noticeable how little difference is made by people's commitments to such general positions about the nature of morality when it comes down to arguing about some concrete moral question.... Thus, everyone proposes general principles, derives more specific principles from them, tests them by examples, argues from case to case by analogy, and so on.

In what follows, I shall make no commitments on meta-ethical matters. Questions like whether moral values belong to the fabric of the universe or whether they are invented will be put to one side. I shall, however, explicate the practice of proposing general principles and deriving and testing specific ones.

Answers to questions of right and wrong may be made on a case-by-case basis. Asked if we should cease or continue activities that risk precipitating climate change, I can confine myself to considering relevant aspects of this particular problem. If asked whether we should cease or continue activities that threaten to destroy tropical forests, I can do the same with respect to that problem. My answers need not be premised on general considerations pertaining to both cases and possibly others – at least not explicitly. If every environmental dilemma is *sui generis* – so complex that next to nothing can be said about such dilemmas in general terms – normative assessment *must* be made on a case-by-case basis. But environmental dilemmas have things in common that may warrant a generalized approach. Thus, they all involve judgements about the relative importance of catering to different people's interests, which indicates the relevance of general assumptions as to how various kinds of interest compare in terms of normative force. Assumptions like this are made all the time when people – philosophers and non-philosophers – grapple with questions of right and wrong. What sets philosophy apart from ordinary thinking on this score is greater care when it comes to elaborating the assumptions in question and more systematic efforts at justifying them.

Very often philosophers take ideas that are thought to be resonant among people in general – the moral conventions of a society, if such exist – as their point of departure. I have already done so myself. Thus, the fact that the search for normative standards to resolve environmental dilemmas begins with an argument about human interests reflects the widespread and pronounced bias towards an anthropocentric view of the environment.[1] I shall continue drawing inspiration from general assumptions that belong to everyday moral thinking, and the reason is both that I find them *prima facie* congenial and that I do not want my arguments to be brushed aside right away as out of tune with how most people think.[2]

Yet, as indicated above, philosophy involves systematic elaboration and justification of ideas; it adds rigour to ordinary thinking and probes the validity of what people commonly think. Vague assumptions are replaced with more precise ones and those that have no rationale are discarded. The tasks of elaboration and justification are, in effect, two sides of the same coin: in order to probe the validity of an assumption, we need a precise statement of it, and if it allows divergent elaborations (as is generally the case), we settle on nothing that does not stand the test of validity.[3]

Granted, then, that we have before us a precise rendering of a *prima facie* congenial assumption as to what generally matters when it comes to resolving environmental dilemmas, how do we find out whether this assumption is justified or not? To be justified, it should ideally be grounded in a moral *theory* – a complete and well rounded argument about the relative importance of different concerns that leads up to the assumption in question. The justificatory force of such a theory lies, as Shelly Kagan (1989: 13) says, in its explanatory function. In his view, an 'adequate justification for a set of principles requires an explanation of those principles – an explanation of why exactly these goals, restrictions, and so on, should be given weight, and not others' (*ibid.*: 13; italics omitted). Only such an explanation can 'help us understand the moral realm'; without it, 'principles will not be free of the taint of arbitrariness' (*ibid.*).

But justification in normative thinking is not over when a *prima facie* congenial assumption has been translated into a precise principle and provided with theoretical underpinnings. It must also, as Barry says, be tested by examples. Such a test consists in applying a general principle to a concrete case and asking whether or not its implication for this case accords with *pre-theoretical intuitions*.

An intuition is a firm conviction – so firm that 'one finds it hard to believe that one could be convinced to give [it] up by any process of argumentation' (Barry 1989: 260). An intuition is pre-theoretical if it comes to me immediately, like a sense impression, independently of theoretical analysis. According to one view, intuitions enter into normative arguments in much the same manner that empirical data enter into hypothetical–deductive tests of descriptive theories. For reasons that soon will become clear, this may be called the argument that *pre-theoretical intuitions have primacy*. The analogy implies, first, that just as a hypothesis is derived from a theory and tested with data to find out if the theory yields correct predictions, an answer to a concrete question can be derived from a general principle and tested with pre-theoretical intuitions to see if the principle comes out right in concrete cases. Second, just as a descriptive theory will have to be revised if its predictions do not fit the facts, a normative principle must be amended if its recommendations do not square with firm convictions about the correct answer to specific questions. A principle cannot stand if it does not accommodate intuitive judgements about concrete cases. Its counterintuitive implications undermine its validity.

A radical objection to the argument that pre-theoretical intuitions have primacy says that intuitions have no justificatory force at all. There are two notable versions of this objection. The first says that an appeal to intuition is an appeal to mere opinion, whether it takes the form of knee-jerk reactions, loose and flimsy ideas or sheer prejudice. The usual reply to this is that a belief does not have the status of a moral intuition unless it is rooted in considered judgements, arrived at through careful, unbiased and well informed reflection. It must be preceded and informed by the fullest reflection that can be bestowed on an act; to paraphrase W. D. Ross (1930: p. 42), convictions are 'evident without any need of proof'. This is not to say that conformism, bigotry or other dubious mechanisms never lie behind strong convictions, but it distinguishes intuition from mere opinion.

The other version of the radical objection to the argument that intuitions have primacy turns on their relativity. One and the same case may, and very often will, be looked upon differently by different persons. However firm a conviction, it represents a personal point of view, coloured by culture and personal character. This cannot be denied, and in philosophical discussion no attempt should be made to conceal the private nature of intuitions. I should appeal to my intuitions about concrete cases when a general principle is tested by examples. But if my intuitions are not out of touch with the way most people see things – if they are neither outlandish nor idiosyncratic – I will not be alone in appreciating their justificatory force.

To conclude, neither version of the radical objection rebuts the argument that pre-theoretical intuitions have primacy. But a more moderate objection has something to be said for it. It focuses on the second aspect of the analogy between hypothetical–deductive tests and tests of normative principles by examples – the notion that normative principles must be discarded or revised if their recommendations are counterintuitive. The question is why a pre-theoretical intuition about a concrete case should be trusted more than the argument leading up to the principle whose validity this intuition might undermine. If the theory has every appearance of plausibility, it arguably lends enough support to a general principle to insulate it against recalcitrant intuitions. Some, like Shelly Kagan (1989), would add that it lends the principle far more support than a few intuitive remonstrances can undo. In his view, a theory-based principle rests on more solid ground than a principle supported only by intuitions, and this line of argument has considerable force. It is not evident that isolated beliefs about specific circumstances provide better guidance than a theory premised on general assumptions about the normative force of different considerations.[4] Pre-theoretical intuitions do not obviously have primacy.

What are we to think, then, if a theoretically sound principle yields counterintuitive answers to specific questions about right and wrong? We might, as John Rawls (1971) says, work back and forth, revising both the principle and our intuitions about concrete cases to arrive at a 'reflective equilibrium' in which moral thought at various levels of generality accord. This is to say that neither theories nor intuitions have unconditional primacy; adjustments must be made and compromises found along the way.

The method of reflective equilibrium imparts some measure of indeterminacy to normative thinking. But neither alternative to this method – the notion that intuitions have primacy or the resolution to disregard intuitions altogether – is much more attractive. Both leave out an important element of moral reflection as we know it from everyday life. Hence, I shall think of justification as a question of attaining reflective equilibrium – working back and forth between a theory of moral priorities and pre-theoretical intuitions.

The theory of interest

It is commonly and reasonably assumed that one ought to be concerned about environmental degradation if it represents a threat to human wellbeing. A person's wellbeing is bound up with the fulfilment of her or his interests,[5] which divide into two categories. People have an interest in some good or state of affairs, first, if they want it or wish it to come about, and, second, if it would be good for them to have or experience it, regardless of whether this is what they want. There are, in other words, interests that reflect personal preferences or desires and interests whose fulfilment actually improves people's wellbeing. Take three examples: the enjoyment of drugs is sometimes (i.e. for some people) in the first category, never in the second; periodic sleep always belongs to both categories; and physical exercise often belongs to the second, not so often to the first.

An important subset of the second category are interests whose nonfulfilment leads to death or serious disease.[6] They relate to the necessary conditions of staying alive and leading a healthy life. I shall call them *vital needs* or, when the meaning is clear, simply needs. The goods that go into their fulfilment will often be called means of *subsistence*. The need for adequate nutrition is among them, as are protection from the elements and medical care in case of illness. Without such goods, people cannot for long function normally.

How many vital needs are there? The list will on no account be long. Most objective conditions of wellbeing do not constitute means of subsistence. This goes, in particular, for what we may call the prerequisites of a decent existence. Those who find no outlet for emotional self-expression or have no opportunity of aesthetic enjoyment lead impoverished lives. The opportunity of exercising these faculties contributes appreciably to a person's wellbeing, but is no vital need, as death or illness do not follow in its absence. Sometimes, moreover, people are said to need elementary education and work. These goods facilitate access to food and health care, yet there is no vital need for them. Food and health care are accessible by other means as well. To be sure, policies that deprive people of income will often, as a matter of fact, deprive them of subsistence, but the necessary conditions of survival and normal biological functioning must not be confused with concrete and replaceable embodiments of these conditions.

Besides vital needs and other interests that refer to objective conditions of human wellbeing, there are, as indicated, interests reflecting personal preferences or desires. Some of the things people desire are objective conditions of wellbeing, too, but many are not and may even undermine physical health. Think of a person who works so hard to fulfil academic ambitions that he is always exhausted and ruins his health. In general, the set of someone's vital needs is not necessarily a subset of her desires, and her desire for something she also needs will not always be gratified by another good that meets her need equally well.

A crucial question is whether or not interests of different kinds can be ranked according to how much their fulfilment matters. Can they, in other words, be compared in terms of *normative force*? Two blunt assumptions may serve to get the discussion started.

> *Premise (1).* The fact that someone has a vital need for something is a weighty reason for seeing to it that he or she gets it.
> *Premise (2).* There is no equally strong reason for seeing to it that anyone's desires are fulfilled.

Before the validity of these premises is examined, consider some logical implications. Two things follow from premise (1). First, there is a *negative* duty not to deprive anyone of what is necessary for survival and normal biological functioning. Second, there is a *positive* duty to assist needy people in meeting vital needs, that is, to do what can be done in order that someone who would otherwise die or suffer serious illness retains life and health. There is no saying, of course, that these duties can never be overridden by countervailing normative considerations. Premise (1) implies only that indifference to how our acts and omissions affect other people's lives and health is morally untenable, not that we are never, all things considered, justified in depriving others of subsistence or failing to assist them in getting it. It is arguable, for example, that no wrong is done in failing to aid a person who has deliberately wasted his means, knowing that this would eventually make him incapable of supporting himself. Considerations of desert might justify doing nothing about unmet needs. But this is not the place to dwell on noninterest-based considerations that might conceivably countervail premise (1). More will be said in chapter 4.

It follows from premise (1) and premise (2) that the fact that vital needs cannot be protected or provided for without dismissing someone's desires is no reason not to protect or provide for them. As the fulfilment of vital needs is more urgent than the gratification of desires, concerns about desires cannot countervail considerations of need. This is to say that the two premises entail a criterion for ranking interests in terms of normative force.

Premise (1) is innocent enough. It does not say that human needs are all that count or that their fulfilment always comes before every other concern, only that problems of subsistence are among the most pressing moral concerns. This might presumably be charged with what Stuart Hampshire (1989: 90) calls the 'constant presuppositions of moral arguments at all times and in all places'.[7]

But premise (2) is controversial. It may be objected that the normative force of needs does not invariably exceed that of desires. Sometimes it does and sometimes it is the other way round. To deflect this objection, consider the distinction between two kinds of interests: some can only be appreciated vicariously, through empathy, by learning the aims and projects of those whose interests we are talking about, while others have an impersonal value which is in the nature of things and makes an immediate impression on other people. The latter goes, for example, for the interest in assuaging severe pain. It can be valued from an objective point of view, and this, according to Thomas Nagel (1986: 166-71), explains why the need for a painkiller has greater normative force than personal preferences for things that merely make life pleasant.

However, James Griffin doubts whether 'the link between need and obligation' is always tighter than that between desire and obligation:

> whatever in the end goes on the list of basic needs, there are likely to be persons who want things off the list more than they want things on it. A group of scholars may, with full understanding, prefer an extension to their library to exercise equipment for their health. And part of what makes us think that basic needs, such as health, are more closely linked to obligation than are desires is that basic needs seem the 'bread' of life and desires mere 'jam'. But an extension to the scholars' library may not seem like 'jam' to them. On the contrary, if the scholars' preference is sufficiently informed then the library is of greater value to them. But then to maintain that needs create obligations where mere desires do not, or that they create stronger obligations, is to say that we have an obligation, or a stronger one, to the scholars to give them what they themselves value less, which would be odd.
>
> (Griffin 1986: 45)

According to Griffin, the view that needs take precedence can be sustained only if desires are mostly morally lightweight, which they are not. In 'the class of mere desires [there are] many heavyweight values – as values go, far weightier in many cases than needs, and weightier in a way that morality cannot ignore' (*ibid.*: 46).

One problem with Griffin's argument is its vagueness. What it invokes as a basic need – the need for exercise equipment – does not obviously fit this label. But we may tighten up the example to make it serve the intended purpose. Imagine a scholar with severe muscular enervation who depends on regular work-outs to function normally. Without exercise equipment, his vital need for health care will go unmet. Imagine, furthermore, that this scholar prefers an extension of his library to exercise equipment. Should we give him what he himself values less because his body would otherwise decay below the level of normal biological functioning? If social security entitles him to a bicycle and he wants to exchange it for books, should his request be accommodated? Or should the offer of assistance be withdrawn if he does not accept the bike? Griffin suggests that the standards of private morality differ from those of political morality in cases like this.

> We ought to distinguish the question, What are the moral standards for distribution? from the question, What are the standards appropriate to distribution on a large social scale by the state? A social worker, confronted with a cripple who asks for a record player of equal value to the wheelchair that the state offers him, will turn him down.... For many reasons, the principles of political and of individual morality are likely to be somewhat different.
>
> (*Ibid.*: 46)

According to this view, public officials should indeed distribute resources on the basis of objective and impersonal criteria as to what recipients need or can do without. But this is a principle of public morality, not a basic moral premise. In private life, Griffin contends, the personal point of view has greater authority. Thus, if 'the cripple is my son, and he says he would much prefer an education in philosophy to lifts and wheelchairs, and I am satisfied that he knows what is involved, I should use my resources as he wants' (*ibid.*: 46).

I can think of two reasons why private but not public beneficence should be attuned to the preferences of intended beneficiaries. The first is that fine-tuning public policy is possible only if authorities collect and analyse vast amounts of intimate information about people's wants and their reasons for wanting what they do. This would be difficult and costly, and might even constitute an intolerable intrusion on private lives; hence, different standards are requisite in personal and public affairs. But the application of personal criteria to public matters need not take this intrusive form. People may themselves be held responsible for reporting private interests and priorities. Then desires that they wish to hide from the authorities will not be eligible for public subsidies, and there is no reason why they should be.

A second reason not to attune welfare policy to personal preferences is that this will drain public funds by inflating claims for assistance. It raises a risk that people cultivate expensive interests and exaggerate their wants which does not arise as long as resources are distributed according to need. In the latter case, there is an objective measure of how much each recipient can claim, but desires are limitless. One may, to be sure, fix the aggregate amount of resources available for distribution, but this solution is incomplete without some rule for sharing out the lot. If it is done according to the intensity of individual preferences, the problem of inflated desires remains. If, by contrast, one stipulates equal individual quotas, the distribution will be insensitive to real differences – caused, for example, by handicaps – between the resources different people require to attain the same level of wellbeing. Some kind of compromise solution can perhaps be found, but the problem still shows that there are practical advantages to basing public policy on general criteria of need.

There is also, however, room for a principled defence of the view that vital needs trump desires – a defence that applies to both private and public affairs. It may be argued that needs always take precedence in cases of *interpersonal* conflict – when, that is, one person's desires can be fulfilled only by letting the vital needs of another go unmet. Thus, if resources may be used either to extend a scholar's library or provide for an ailing person's health, and if scarcity bars both options, then the sick person's need comes before the scholar's desire. Moreover, if the scholar is also in need of medical aid but intends to swop medicine for books, then resources are better spent on an equally needy person who cares more for health. The scholar may end up very unhappy as a result, but unhappiness is the lesser evil when life and health are at stake.

The last argument suggests a solution in two parts to the problem of ranking interests according to normative force. On the one hand, personal preferences may well be decisive as far as *intrapersonal* conflict is concerned. While there may be practical obstacles to basing public policy on this criterion, there is in principle no reason to deny people what they themselves value most as long as the opportunity cost of providing for their desires rather than their needs falls on them alone. On the other hand, when it comes to interpersonal conflict, the opportunity costs of giving one person what she wants fall on others whose interests will, to a greater or lesser extent, be set aside. Then the distinction between objective needs and subjective preferences takes on importance, and there is good reason to be most concerned about those whose survival or normal biological functioning is at stake. No one should have to undergo death or serious physical suffering in order that others may be able to gratify their desires.

Hence, premise (2) is a justified appendage to premise (1). There is a duty not to do anything that makes anyone incapable of meeting vital needs if the adverse effect of doing something else is only to make someone incapable of fulfilling their desires. Desires that can be gratified only by dint of depriving someone of the means subsistence have to be renounced.

The argument above will be referred to as the *theory of interest*. It offers a basis for working out normative guidelines for environmental policy. To be sure, such guidelines should not be worked out solely on this basis, and I shall later (chapters 4 and 6) consider arguments calling for a broader outlook. But the theory of interest will remain central throughout the book.

As it stands, the theory of interest entails both a negative duty of not depriving people of what they need to meet vital needs, and a positive duty of assisting them in getting it. It may be objected that this is too much. There exists a negative duty of nonmaleficence – so the objection goes – but no positive duty of beneficence. This echoes common sentiments. Even those who countenance positive duties tend to speak of them in weak terms. Witness James Fishkin: 'If a person knows he can prevent *great* harm, such as the loss of a human life, he is morally obligated to do so if the costs to him (and to anyone else) are *minor*' (Fishkin 1982: 65; italics added). But there are problems with the deflation of positive duties, and the distinction between negative and positive duties may not make sense.[8] Fortunately, I can defer taking sides in this debate. The status of the duty of beneficence is almost totally irrelevant to what will be said below about the goals of environmental policy. Not until the final paragraphs of the last chapter will the discussion go beyond the negative duty of nonmaleficence. Doubts about beneficence may therefore be all but neutralized by replacing premise (1) with:

> *Premise (1').* The fact that someone has a vital need for something is a weighty reason not to deprive him or her of it.

The priority principle

The next task is to translate the theory of interest into a guideline for environmental policy. We need to know the practical implications of the assumption that vital needs rank above desires as far as normative force is concerned. How are political priorities to be defined in concrete cases of environmental policy making? I shall develop a *priority principle* to guide us on this score. Its contents will be brought out in two steps. First, an extreme principle that accords absolute priority to protecting vital needs is put forth. Then a critical discussion of this principle leads to the formulation of a moderate and more plausible principle.

Let me begin, then, by stating the strongest normative principle that might possibly be derived from the theory of interest:

> *The extreme principle.* It is imperative not to engage in activities whose impact on the natural environment deprives or risks depriving anyone of the means of meeting vital needs.

This principle implies that it is wrong to do anything that has any chance of taking away anyone's means of meeting vital needs when the cost of not doing so is only to leave someone's desires unfulfilled. One must not jeopardize any person's life or health in order that other people, however many, will be able to get what they want but do not require for survival or normal biological functioning. Vital needs enjoy lexical priority.[9]

A problem with the extreme principle immediately comes to mind. If the desires of a large number of people will be sacrificed to the needs of a much smaller number – perhaps just one person – we may have second thoughts about abiding by the principle. To have qualms on this score is not tantamount to disregarding the different normative force of needs and desires. It rather reflects a hunch that normative assessment should be desire-sensitive, as it were. The greater importance of meeting needs is acknowledged, but so is the existence of a limit to how *many* lightweight interests one can set aside for the sake of someone's subsistence. The question whether such a limit exists or not is one aspect of what I shall call the problem of *desire-sensitivity*.

Qualms may also be had about the fact that the extreme principle puts actions that are certain to deprive someone of subsistence on a par with actions that only risk doing so. It is arguable that the risk of an adverse outcome is not as bad as a certain adversity. The badness of the former should perhaps be reduced by a factor corresponding to the prospect that no disaster comes about. If so, the assessment of an action that only risks causing disaster must be adjusted to take account of the chance that everything goes well. This issue brings up the second aspect of desire-sensitivity. Assume that doing A may undermine the subsistence of some people, while not doing A is certain to make other people incapable of fulfilling their desires. The extreme principle rules out A, but one may wonder if the badness of this action is to be reduced on account of the uncertainty over its adverse effects. If it is, the less harmful but certain outcome of not doing A may appear to be as bad as, or even worse than, the more harmful but uncertain outcome of A.

Anything that calls the extreme principle into question may lower the moral barrier against policies and practices which put subsistence at stake. There may be other problems with this principle besides desire-sensitivity, but I shall not provide a long list of potentially countervailing considerations here and now. Some will be confronted along the way, but, at this stage, none are as intrusive as the problem of desire-sensitivity. It derives from questions that immediately spring to mind when the moral importance of human wellbeing is appreciated: how many people will prosper or suffer if we do one thing rather than another, and how likely are they to prosper or suffer?

There are two conceivable solutions to the problem of desire-sensitivity. The first corresponds to the extreme principle: we may insist that considerations of need always constitute conclusive reasons for action. The alternative solution is that such considerations sometimes give way to desires. Which is correct?

Let me begin with the first aspect of desire-sensitivity. A hypothetical case may serve to focus our attention.[10]

> *The case of S.* The political authorities of society S contemplate a large cut in public expenditure on pollution abatement measures. It is aimed at reducing a heavy burden of taxation, thereby improving people's purchasing power and, hence, increasing their opportunity of gratifying many desires. But there is a risk, albeit a small one, that fewer abatement measures will lead to leakages of certain toxic gases, and there is also a small risk some people will die from exposure to these gases. This is to say that lives which will be saved if public expenditures are not cut may be lost if they are cut. Should a cut be made?

The extreme principle says no, but an affirmative answer seems more plausible. Granted, more specifically, that (i) many people of S will benefit greatly from a cut in costly abatement measures, (ii) it is the least risky route to the benefits in question, (iii) there is only a small risk of fatal pollution, and (iv) nothing besides this risk speaks against the cut, it should be made – or so, at any rate, I believe. A small chance that vital needs will go unmet is a lesser evil than a certain opportunity of enabling many people to get what they want. It seems sensible to live with such a risk if this is the only way of achieving substantial welfare improvements. In every society, many decisions and practices bespeak such a view. Thus, cuts in expenditure on public health are regularly being countenanced even if the concomitant reductions in health care can have fatal consequences. The measures contemplated in S are, I think, equally justified. The implication of the extreme principle with respect to this case fails to square with my pre-theoretical intuition about its correct handling.

What does this observation imply with respect to the validity of the principle? An intuition is a firm conviction – so firm, to recall the definition quoted above, that 'one finds it hard to believe that one could be convinced to give [it] up by any process of argumentation' (Barry 1989: 260). Thus, I steadfastly doubt that I could be talked into denying that the measures contemplated in S are justified. Moreover, my intuition about S is pre-theoretical – it is not arrived at through an evaluation of the case in the light of general principles of right and wrong, but comes to me immediately.

Now, the extreme principle was motivated by a theory about the normative force of different human interests. This theory has every appearance of plausibility, and it is arguable, as I said under 'Method', that a theory may lend enough support to a principle to insulate it against recalcitrant intuitions. But this line of argument can salvage the extreme principle only if the principle is derived from and, hence, explained by the theory of interest. This, however, is not the case. The extreme principle is not entailed by the argument that vital needs enjoy greater normative force than desires. The theory of interest neither implies nor rules out that people ought to live with a small risk of death or physical suffering in order to enhance the opportunity of gratifying desires. The extreme principle is, in other words, compatible with the theory, but no logical consequence of it. In view of this, my pre-theoretical intuition about the case of S supplements rather than challenges the theory of interest.[11]

Thus, unless we believe that pre-theoretical intuitions have no justificatory power at all and cannot decide moral issues even in default of theoretical considerations, we must, after confronting the extreme principle with the case of S, conclude that the principle blurs a morally relevant distinction between certainty and risk. An action that is certain to make someone incapable of fulfilling desires may be worse than an action that risks depriving someone of the means of meeting vital needs. It is, accordingly, not always wrong to impose such a risk on a group of people if it can be averted only at the expense of another group's more lightweight interest. Something must therefore come in place of the extreme principle – a guideline of normative assessment that both has a basis in the theory of interest and displays desire-sensitivity.

Before more is said about the problem brought out by the case of S, I shall explore the second aspect of desire-sensitivity to find out if this, too, calls for a less extreme guideline for environmental policy. Can considerations of need ever be outweighed by a concern about desires if death or serious deprivation are certain to ensue should needs receive less than absolute priority? To test our intuitions on this score, new imaginary cases are required:

> *The case of S'*. The political authorities of society S' contemplate a cut in public expenditure on pollution abatement. Their aim is welfare improvements of the same magnitude that a similar policy will effect in society S. In S', however, the contemplated cut is certain to have fatal consequences. With reduced abatement measures, there is no way a certain chemical substance can be prevented from leaking out, and no chance of protecting the general population from exposure to it. While most people will not be adversely affected by such exposure, it is known to have deadly effects on a small group with a rare and incurable disease. Should the contemplated cut in public expenditure be made?

According to the extreme principle, it should not. If those who suffer from the rare disease go unprotected, their vital needs will be sacrificed to other people's desires. This is a certain effect of cutting public expenditure on pollution abatement.

How does the negative answer to the case of S' square with pre-theoretical intuitions? Much better, I think, than the negative answer to the case of S. No magnitude of increase in nonessential welfare – however high and however widely distributed – can offset the moral evil of certain death or sustained and severe physical suffering. This is my firm conviction. There is, accordingly, nothing implausible about the implication of the extreme principle with respect to S'. Unlike the case of S, it gives no reason for replacing the extreme principle with a desire-sensitive guideline of environmental policy. But consider another case:

> *The case of S"*. The authorities of society S" contemplate a cut in public expenditure on pollution abatement measures. As in S and S', the purpose is a substantial welfare improvement in the form of a much lower level of taxation. If the cut is made, a small number of people are virtually certain to die owing to exposure to toxic waste which can no longer be stored safely, but we cannot tell in advance who they will be.[12] Should the cut be made?

Once more, the extreme principle yields a negative answer. Is this intuitively correct? The cases of S' and S" are largely alike, the only difference being that we know whose lives are at stake in S', but know only that there are lives at stake in S". This difference may, however, incline some who oppose the cut contemplated in S' to favour the cut in S". We react, as Jonathan Glover (1977: 210) says, differently to the death of a known person than to 'statistical' death. As an illustration, he points to 'the discrepancy between what society will sometimes spend on saving the life of a known person in peril and what it will spend to reduce the future level of fatal accidents'. Thus, when large resources are about to be spent on rescuing a trapped coal miner, no one proposes that the rescue operation be halted and the money diverted to precautionary measures with the potential of preventing many similar accidents in the long run. Glover explains this by the 'extreme horror which most of us have at the thought of being in a position where we can see certain death ahead' (*ibid.*: 213). The previous remarks suggest a more precise explanation: it is the extreme horror of, as it were, condemning a *known* person to death that renders the current rescue of a trapped miner more urgent than the prevention of many future accidents. Such horror also attaches to the policy that is contemplated in S', but not to the one under contemplation in S", whose prospective victims are anonymous.

It is hard to tell how much of the horror of certain death that anonymity takes away. I, for one, profess no firm conviction about the correct handling of case S". It is significant, then, that the theory of interest does not sanction the policy contemplated in S". Nothing about the normative force of a vital need hinges on the identity of the one whose need it is. A certain but anonymous death is no less an evil than the certain death of a known person. So, all in all, consulting intuitions as well as theoretical considerations, we find that the case of S", like that of S' but unlike that of S, brings out no weighty reason for revising the extreme principle.

One formidable objection remains: even if neither theoretical assumptions nor pre-theoretical intuitions permit differential treatment of anonymous deaths and the deaths of known people, the practical implications of ruling in favour of environmental protection in the case of S" reduces the argument to absurdity. Shifting from environmental policy to the transport sector, for example, this argument forces us to conclude that we should not use cars, planes, trains or other means of transportation that are certain to cause fatal accidents once in a while. Similar absurdities may easily be pointed out in other areas. They show, according to the objection, that anonymous deaths and the deaths of known people can be put on a par only on pain of recasting modern civilization.

It is clearly correct that political practice everywhere bespeaks a willingness to see certain but anonymous death without taking every available precaution against it. Public policy is nowhere being shaped with a view to maximizing the number of citizens who enjoy the highest possible chance of survival – through better road safety, improved health care, enhanced workplace security, and so on. However, as long as we base our judgement on a comparison between the relative urgency of meeting needs and gratifying desires, the morality of the matter is, at the very least, a moot point. If a radical break with present practice really is absurd, it must be because things other than this comparison bear heavily on the matter. The theory of interest need not exhaust all normative considerations that are relevant to decisions about public policy. I shall consider other matters in chapters 4 and 6.

This concludes the discussion of desire-sensitivity. It tends to confirm the categorical ban on activities that are certain to deprive someone of the means of meeting vital needs, but loses the ban on risky activities. The extreme principle must therefore be replaced by a more flexible guideline for environmental policy.

> *The priority principle.* It is imperative not to engage in activities whose impact on the natural environment deprives or risks depriving someone of the means of meeting vital needs, unless the risk is too small to be a reasonable source of concern.

The last clause of the priority principle leaves a large question mark: how are we to distinguish risks that are reasonable sources of concern from those that are not? Or, as I shall henceforth say, where does the line between *significant* and *insignificant* risk go? This is a major subject of the next chapter, which further elaborates the priority principle before it goes on to consider environmental dilemmas that take us beyond this principle.

Notes

1 The bias is clearly stated by the World Commission on Environment and Development: the 'environment does not exist as a sphere separate from human actions, ambitions, and needs, and attempts to define it in isolation

from human concerns have given the very word 'environment' a connotation of naivety in some political circles' (WCED 1987: xi).

2 The last motivation is discussed in Malnes (1992).

3 While philosophers are commonly critical of conventional ideas, they rarely agree on right and wrong, and this may throw doubt on philosophy as a tool for justifying general assumptions that enter into normative thinking. Perhaps philosophers can do no more than putting ideas straight and clearing the way for further debate. It may seem, indeed, that the limits to conclusive justification are narrow, but this is not to say that philosophical inquiry makes no difference when it proves inconclusive. It may throw doubt on the intellectual credentials of certain opinions and place a burden of proof on those who hold them, which is no trivial matter to people who set store by rational justification.

4 Kagan (1989: 12) buttresses his argument by pointing out that pre-theoretical intuitions may reflect theoretical assumptions and questioning whether intuitions exist at all: 'It would be more correct to say that we begin moral philosophy already possessing some moral theory – albeit theory which is only half-formed, and largely unarticulated.' But the fact that convictions about concrete cases are charged with theoretical assumptions no more implies that such convictions are indistinguishable from theories than that the theory-loaded character of empirical data renders these indistinguishable from descriptive theories.

5 Yet 'interest' has a wider conceptual scope than 'wellbeing'. Interests relate to everything a person cares about or is affected by, while wellbeing depends only on things that make that person better or worse off. Thus, 'An agent may take an interest in the welfare of another person ... even though he does not expect that person's well-being to lead to a more immediately personal benefit for the agent himself' (Kagan 1989: 234). Still, an increase in a person's wellbeing presupposes an increase in the fulfilment of that person's interests.

6 Some interests have an uncertain status in that their nonfulfilment is catastrophic for some people, but not for others. Consider, for example, the interest of a newborn child in contact and communication with other persons. (I owe this example to Jon Hovi.) There is a risk that a person will die if left too much alone in infancy. However, the strength of the need for contact differs between individuals, some being less vulnerable to deprivation on this score than others. Such variation exists to a far lesser extent with respect to the needs of nutrition and medical care. Hunger kills everyone after a while, and the same goes for certain illnesses. The needs for rest and protection from the elements fall somewhere between. Everyone will pass away from excessive exhaustion or freeze to death if their body temperature falls too low, but some stand strain and extreme conditions better than others.

7 Hampshire (1989: 90) writes: 'That destruction of human life, suffering ... are, taken by themselves, great evils, and that they are evil without qualification, if nothing can be said about consequences which might palliate the evil ... – these are some of the constancies of human experience and feeling presupposed as the background to moral judgments and arguments'.

8 For an effort to discredit the distinction, see Harris (1974). For an overview of philosophical and juridical opinions, see Goodin (1985: 18–23).

9 Lexical – or lexicographical – priority is the principle behind the order of words in a dictionary. Every word that begins with an earlier letter in the alphabet is listed before words that begin with later letters; if two words begin with the same letter, priority goes to the one whose second letter comes first, and so on. The concept gained currency in political theory through John Rawls' (1971: 42–4) use of it.

10 Those who are unfamiliar with analytical philosophy may find the resort to hypothetical cases strange. Why not talk about events from real life? The reason is explained well by F. M. Kamm (1993: 7): 'Real-life cases often do not contain the relevant – or solely the relevant – characteristics to help in our search for principles. If our aim is to discover the relative weight of, say, two factors, we should consider cases that involve only these two factors, perhaps artificially, rather than distract ourselves with other factors and options.'

11 Falling back on intuitions may be premature before we have inquired whether or not the argument for the theory of interest can be extended to provide adequate justification of the extreme principle. Well, can it? Possibly, but I do not know how. Nor, incidentally, do I have a theoretical argument to offer in support of my intuitive remonstrances against the extreme principle. Hence, no explanation is at hand for either rejecting or countenancing the measures contemplated by the authorities of S. Intuitive dissatisfaction can, accordingly, 'tell us that something is wrong, without necessarily telling us how to fix it,' which is to say that 'our moral understanding extends farther than our capacity to spell out the principles that underlie it' (Nagel 1991: 7).

12 This situation is not uncommon. 'In many cases, the risk may be well understood in a statistical sense but still be uncertain at the level of individual events. Insurance companies cannot predict whether any single driver will be killed or injured in an accident, even though they can estimate the annual number of crash-related deaths and injuries in the U.S. with considerable precision' (Morgan 1993: 33).

3
Risk and hard cases

Introduction

This chapter continues what chapter 2 began: the construction of a normative standard that makes the resolution of environmental dilemmas a matter of meeting the most urgent interests involved. I shall first elaborate the *priority principle* and then extend the argument behind it to deal with dilemmas that this principle does not resolve. When these tasks are accomplished, we have in hand a complete set of interest-based criteria for defining the goals of environmental policy.

The elaboration of the priority principle is needed first, to distinguish between two types of risk associated with activities that may deprive people of the means of meeting vital needs: risks that are reasonable sources of concern, and risks that are not. Next, under 'Needs in conflict', the argument behind the priority principle is extended. This section asks what ought to be done if every available course of action deprives someone of subsistence or has a significant risk of doing so. The aim, I believe, should then be to minimize the extent of death and suffering. The next section asks how an environmental dilemma ought to be handled if no alternative is a significant threat to anyone's vital needs. I answer that the preservation of pristine nature has particular importance in a modern, industrialized society, and should be reflected in procedural limits on democratic decisions, with analogy to constitutional constraints protecting civil rights.

Modes of ignorance and risk evaluation

The priority principle says that it is imperative not to engage in activities whose impact on the natural environment deprives or risks depriving someone of the means of meeting vital needs, unless the risk is insignificant, that is, too small to be a reasonable source of concern. Where does the line between significant and insignificant risk go? I shall argue that the proper way of drawing it varies with the circumstances, depending on what basis we have for *estimating* risks.

A risky action is one whose outcome depends on circumstances beyond our knowledge and control. Let A_i be a risky action and assume that the unknown circumstances are in state S_1 or S_2 ... or S_n. To each pair of act and state, A_i and S_j, there corresponds an outcome, O_{ij}. If A_i is chosen and it turns out that things are in state S_1, O_{i1} will obtain. Thus, the risky action 'defines a consequence as a function of the state of the world' (Arrow 1966: 254). This will be called a situation of *ignorance*. It may be depicted as follows:

	States of the world			
	S_1	S_2	S_j	S_n
Act	*Outcome*			
A_i	O_{i1}	O_{i2}	O_{ij}	O_{in}

In this situation, the true state of the world is unknown but nothing else is. First, we know the range of possible states of the world. S_1 ... S_n is an exhaustive enumeration of everything that may conceivably be the case with respect to external circumstances. Second, the nature of all outcomes is supposed to be known. For every S_j, we can say what will happen if the risky action is done and things happen to be in this particular state. Each O_{ij} describes a consequence, and 'in the description of a consequence is included all that the agent values, so that he will be indifferent between two actions which yield the same consequence for each state of the world' (Arrow 1966: 254). Moreover, the fact that we do not know the true state of the world need not imply that we are totally ignorant even on this score. Ignorance has several modes, and for each mode I shall now suggest how the distinction between significant and insignificant risk can be drawn so as to determine whether or not various environmental risks are reasonable sources of concern.

Numerically estimable risk

This is the least serious mode of ignorance. We know enough to assign numerical probabilities to possible states of the world. Thus, there corresponds a probability estimate, p_j, to each S_j, such that $0 \le p_j \le 1$, and $\Sigma p_j = 1$. Similar estimates can be assigned to the possible outcomes of the risky action. If A_i is chosen, the probability that O_{ij} results is equal to the probability of S_j.

How can probabilities be known? First, they may be estimated by means of frequency analysis. If a situation repeats itself with one out of a fixed number of states arising on each repetition – if, for example, a coin is tossed consecutively to see if it comes out heads or tails – the likelihood of a given state equals its relative frequency. Frequency analysis is, in particular, a common method for estimating the likelihoods of certain illnesses in different population groups. Probabilities deriving from such analysis are *objective*.

Second, probabilities may be estimated by means of a scientific theory. This is done by geologists calculating the chance that a major earthquake will take place in California during the next five years. What they estimate is, in Jon Elster's (1983: 196) terminology, *theoretical probabilities*. Theories can, of course, be premised on frequency analysis, but theory-based calculations may yield probability estimates of events which have never occurred and may never actually take place. Thus, under 'the classical theory of statistical mechanics, a beaker of water has an exceedingly small but nonzero chance of freezing spontaneously' – a probability 'so small that such an event is unlikely to happen in the lifetime of the universe, although given the relevant laws, it is nomologically possible' (Humphreys 1989: 14).

Frequency analysis and dependable scientific theories are straightforward bases for estimating the likelihoods of different states of the world in numerical terms. A more problematic route to such estimates is indicated in the following example.

> During the thirteen days that encompassed the so-called Cuban Missile Crisis in 1962, estimates were offered of probabilities that military intervention by the U.S. in Cuba would trigger a nuclear war with the Soviet Union. Some members of the U.S. National Security Council or of the Chiefs of Staff estimated the probability as 0.3, some as 0.5, etc. And these estimates were offered ... in implicit attempts to cast the problem into a decision under [numerically estimable] risk framework.
>
> (Rapoport 1980: 50)

There is no way these probabilities could have been derived from frequency analysis of similar crises or a theory of international conflict escalation. They must have been based on educated judgement grounded in military and political experience and possibly supplemented by hypotheses and findings from psychology and political science. It is a question of *subjective probabilities,* which largely reflect the particular beliefs and biases of those who estimate them.[1] The credibility of subjective probabilities is, however, a moot question, to which I return below.

Assume, then, that we stand before the choice of pursuing or not pursuing a policy whose impact on the natural environment risks depriving someone of the means of meeting vital needs. Suppose, moreover, that we have a reliable basis for estimating this risk in numerical terms. How are we to tell whether it is significant or not – whether, in other words, the risk associated with a given policy is a source of reasonable concern about its consequences? This may be done by fixing a critical level, *C*, defined in numerical terms, to distinguish significant from insignificant risk. If the risk of depriving people of the means of meeting vital needs is at or above this level, it is significant; lower risks are no source of concern.

Where should C be placed? Here is one proposal: a significant risk exists if and only if the probability of someone being left without the means of meeting vital needs is 10% or more. There is, of course, no way of proving that 10% is the proper cut-off point. Will we endanger people's lives and health by deeming risks below this level insignificant? And do we show insufficient sensitivity to desires by tolerating no risk above 10%? I am inclined to answer both questions in the negative, but cannot offer anything remotely resembling a philosophical rationale in defence of my hunch. Nor, however, will things be easier if we defer making explicit assumptions on this score. To make the priority principle operational, a distinction must be drawn between significant and insignificant risk, and in situations of numerically estimable risk, this is naturally done by specifying the critical level in numerical terms. But there is no denying the debatable and revisable character of any value assigned to C.

The following objection may be raised to the use of critical levels: if C = 10%, a 10% probability of one person's death is worse than a 9% probability of 100 deaths, and this is counterintuitive. It is, moreover, senseless to say that the first risk ought to be averted even if many people will thereby have to renounce things they desire, while it is permissible to live with the second risk in order that desires can be gratified. A similar objection may, of course, be made about any given level of *C*.

The objection rests on the premise that $X + 1$ deaths are worse than X deaths. This, I believe, is a valid assumption which will be defended below (see under 'Needs in conflict') as a crucial basis for dealing with so-called hard cases. Moreover, the objection to critical levels implies that assessments of risk should be premised on comparisons of *expected deaths* rather than estimates of the confidence one has in expecting that anyone will die at all. Expected deaths are most reasonably calculated by multiplying the lives at stake in a given situation with the probability that they will be lost. By this kind of calculation, a 1% probability of 100 deaths is worse than a 99% probability of one person's death. Is this assumption less counterintuitive than the previously stated implication of assessing numerically estimable risk by fixing a critical level? In my view, it is not, but I concede that intuitions diverge on this score.

As a compromise between conflicting intuitions, I suggest the following formula. In order to distinguish between significant and insignificant risk in situations of numerically estimable risk, a critical level must be fixed, but it cannot be fixed without regard to the number of people whose lives are at stake. The higher the number of people whose means of meeting vital needs may be taken away if we engage in a given activity, the lower the level should be set. Start by fixing the level at 10% if life may be lost at all, and adjust it downwards as potential tolls increase. The rate of adjustment cannot be stated precisely, but the formula goes a long way towards narrowing down the room for discretion in assessments of numerically estimable risk.

First-degree uncertainty

If risks cannot be estimated numerically, a situation of *uncertainty* is before us. The true state of the world is unknown and so is the exact likelihood that it corresponds to one of a set of possible states. In such a situation, we cannot distinguish between significant and insignificant risk by fixing a critical level. What should we do?

Before this question is answered, we must face the challenge that uncertainty is not a distinct mode of ignorance, but really nothing but a particular instance of numerically estimable risk. The argument, advanced by the *Bayesian* school of decision theory, says that any situation of ignorance permits numerical estimates of probabilities, not always derivable from frequency analysis or scientific theories, but anyway capable of justification by *subjective* judgement. John C. Harsanyi (1975: 599) argues:

> Bayesian decision theory shows by rigorous mathematical arguments that any decision maker whose behavior is consistent with a few – very compelling – rationality postulates simply *cannot help* acting *as if* he used subjective probabilities.... I shall quote only two of these rationality postulates: 'If you prefer *A* to *B*, and prefer *B* to *C*, then consistency requires that you should also prefer *A* to *C*'; (2) 'You are better off if you are offered a *more valuable* prize with a given probability, than if you are offered a *less valuable* prize with the same probability.' The other rationality postulates of Bayesian theory are somewhat more technical, but are equally compelling.

We may safely take Harsanyi at his word when he says that the rationality postulates are very compelling. The problem is, however, that a person who satisfies these postulates will end up assigning subjective probabilities to possible states of the world only after she has gone through a fairly artificial exercise. It consists of choices between hypothetical lotteries. Thus, to elicit your estimate of the probability that a military coup will take place in Russia during the next five years, I may offer you a choice between two bets: *either* you receive $100 if there is a coup and nothing if no coup takes place, *or* you receive $100 if no coup takes place and nothing if there is one. If you prefer the first bet, it follows that you believe the probability of a military coup in Russia during the next five years to be 0.5 or higher; if you prefer the second bet, you must believe that the coup has a probability of 0.5 or lower. Continuing this exercise by offering you more choices between hypothetical bets, I can, by registering your preferences, arrive at ever more precise estimates of your estimates of the likelihoods that a coup will take place or not. If asked directly, you may be incapable of assigning probabilities to the two states, but the hypothetical lotteries offer an indirect method of explicating your implicit assumptions. The problem is only, as I said, that the method is very artificial. Probability estimates elicited in this manner may not reflect considered judgements about relative likelihoods. Thus, choosing between hypothetical bets, a person may commit himself to consistent numerical assessments, but be unable to patch up a coherent explanation of them. Peculiarities with respect to how the questions are posed may influence responses, and choices may be made at random: being unable to determine which alternative is best, she simply picks one. It is therefore doubtful whether subjective probability estimates can generally serve as premises of rational decision making. Would it be prudent, say, to base official policy towards Russia on predictions of its future arrived at by presenting public officials with hypothetical bets?

The rejection of subjective probability estimates must not be too categorical, however. We should not take an all-or-nothing attitude, thinking that subjective probabilities can enter into rational deliberation only if they come in numerical form, and concluding that they have to be left out because they cannot take this form. There are, as will soon be seen, different degrees of uncertainty about the probability distribution over possible states of the world.[2]

Under what I shall call *first-degree* uncertainty, estimates of probability are determinate enough to make *ordinal* comparisons of likelihoods feasible. States can, in other words, be ranked in terms of how likely they are. Thus, I may know enough about my colleague's habits to say that he is more likely to be in his office than at the pub at this hour, but not enough to fix numerical estimates of either state. Such an ordinal ranking of likelihoods is *complete* if, for each pair of probabilities, p_j and p_k, we can say whether $p_j \geq p_k$ or $p_j < p_k$. It is *consistent* where, for any three probabilities, p_j, p_k and p_l, if $p_j \geq p_k$ and $p_k \geq p_l$, then $p_j \geq p_l$.

An interesting hybrid of numerically estimable risk and first-degree uncertainty is the situation in which ordinal probabilities exist and we know *something* about the differences between them, but not enough to arrive at numerical estimates. I may, for example, be certain that S_1 is much more likely to be the true state of the world than S_2, while S_2 is only slightly more likely than S_3. Such knowledge can be had about the differences between some or all likelihoods. Thus, knowing what was just suggested about p_1, p_2 and p_3 is compatible with not knowing anything about the relationship between p_3 and p_4 except that p_3 is greater than p_4.[3] First-degree uncertainty with this much knowledge of probability differences will be called *mitigated* first-degree uncertainty.

How are we to draw the line between significant and insignificant risk under first-degree uncertainty? I shall propose two rules (the second pertaining only to mitigated first-degree uncertainty).

(i) If an activity is more likely than not to deprive someone of the means of meeting vital needs, it raises a significant risk. If, by contrast, it is more likely that no one will be left without the means of meeting vital needs, the risk is insignificant.

It might be objected that the risk of some activity undermining someone's subsistence may be as high as 49% without exceeding the likelihood that things go well, and, at this level, it should certainly be a source of concern. Under first-degree uncertainty, however, we cannot gauge the absolute levels of risk and the objection is therefore inapposite. It is never known how large or small the difference between the likelihoods of two states of affairs is, only that one is more likely than the other. Hence, the only alternatives to the suggested criterion for identifying significant risk are to say either any or no risk of depriving someone of the means of meeting vital needs is significant.

The second rule pertains to situations of mitigated first-degree uncertainty.

(ii) If an activity is more likely than not to deprive someone of the means of meeting vital needs, or if it is less likely but the likelihoods of the two outcomes are not far apart, the risk associated with this activity is significant.

There is no use specifying how much likelihoods can differ before the risk is deemed insignificant. Estimates on this score will anyway be imprecise under uncertainty, and we have to trust our good judgement in particular cases.

Second-degree uncertainty

If we are ignorant about the true state of the world and totally incapable of estimating the likelihoods of possible states, we face what may be called *second-degree uncertainty*. It leaves us altogether in the dark when it comes to estimating risk. Here is a simple example: I find a mushroom that I have never before tasted and cannot identify; it may prove palatable, unpalatable, or even toxic, and I have not got an inkling as to which condition is the more likely one and what will happen if I eat the mushroom.

Neither absolute nor comparative estimates of probability can be invoked to distinguish between significant and insignificant risk under second-degree uncertainty. There is only information about the nature of possible outcomes to go by. This implies that the last clause of the priority principle cannot be given a precise, operational rendering. Faced with the choice of pursuing or not pursuing a policy that risks depriving someone of the means of meeting vital needs, we have no way of telling whether the risk is too small to be a reasonable source of concern. This leaves two options: either disregarding risk in every case of second-degree uncertainty or considering all such risks significant. In view of how badly people will fare if worst comes to worst, the first option is out of the question. It follows that a risk that someone will be left without the means of meeting vital needs is always a reasonable source of concern under second-degree uncertainty.

It may be questioned, however, whether this mode of ignorance has much empirical interest. We are arguably never without a basis for any kind of judgement about relative likelihoods. Thus, in the mushroom example cited above, I should have concluded from the relative rarity of toxic mushrooms that the one I found is least likely to be of this kind.

To be sure, if we are ignorant not only about the actual state of the world, but even about the *range* of possible states, ordinal probability estimates are unavailable. Then the world may be in states that lie entirely beyond our imagination and which cannot be considered in advance. It has turned out, for example, that some illnesses which eventually arose from exposure to modern chemicals were not among the prefigured hazards of these substances, which is to say that the choice between using or not using the chemicals was not made against the background of a well defined set of possible outcomes. In such a situation, there is no reliable basis for ordinal estimates of probability. As things may happen that will come as a surprise to us, we are likewise in for surprise with respect to the likelihoods of known contingencies.

Robert Goodin (1978: 35) explains the distinctive feature of this kind of ignorance, which he calls *profound* uncertainty: 'we might be uncertain of the future because we cannot even imagine all that could possibly happen. We are prepared to be surprised not just by outcomes that we thought unlikely but, more fundamentally, by outcomes we had not thought about at all.' A telling problem emerges, however, in Goodin's further elaboration of profound uncertainty. Alluding to the method of fault tree analysis of nuclear reactors, he says that certain 'classes of question simply are not amenable to fault tree analysis precisely because it is in principle impossible to list all paths leading to failure.... Consider the danger of sabotage of the nuclear reactor: were we able to anticipate all paths would-be saboteurs might take, we could block them and eliminate the danger altogether' (*ibid.*: 36). It is not clear, however, if this is an example of profound uncertainty. It may better be described as a situation in which ignorance of probabilities is combined with full knowledge of possible states of the world. We start from a set of possible states of the world which include the occurrence of sabotage, but are unable to estimate the probability of this particular state because it may come about in more ways than we can think of. (Countenancing Goodin's classification would be tantamount to saying that almost every choice involving advanced technology is made under profound uncertainty.[4]) But we are rarely incapable of saying anything at all about the risk that a nuclear reactor will be sabotaged. Depending on the amount and quality of security measures, there will typically be a sound basis for comparing the likelihood that would-be saboteurs succeed with their chance of failing. In general, many situations that appear like cases of profound uncertainty can on closer scrutiny be categorized as first-degree uncertainty.

Still, the example of modern chemicals shows that profound – or second-degree – uncertainty is not an empirically void category. The limits of imagination may leave us with an incomplete specification of possible states. To be sure, if things we did not prefigure come as a complete surprise, we had no basis for assuming in advance that our choice was made under profound uncertainty. But there exist situations in which one may reasonably conjecture that unforeseeable outcomes are among the possible ones. This is presumably the case after an established political order has broken down and an array of social forces been let loose.[5] I doubt, however, that this kind of situation is common.

Summary

We now have four criteria for distinguishing between significant and insignificant risk:

(i) In a situation of numerically estimable risk, a critical level must be fixed. It might be set at 10% if any life may be lost and adjusted downwards as potential tolls increase.

(ii) Under first-degree uncertainty, any activity that is more likely than not to deprive someone of the means of subsistence raises a significant risk.

(iii) Under mitigated first-degree uncertainty, if an activity is more likely than not to deprive someone of the means of subsistence, or if it is less likely to do so but not much, the risk associated with this activity is significant.

(iv) Under second-degree uncertainty, a risk that an activity will leave someone without the means of subsidence is always significant.

Needs in conflict

What if every available policy will either leave someone without the means of meeting vital needs or raise a risk to this effect? What, in particular, if both the destruction and the protection of the natural environment represent a threat to subsistence? Then we are confronted with a hard case that cannot be resolved by invoking the priority principle, which does not say what ought to be done if the effect of not inflicting damage on the environment is also to deprive or risk depriving people of what they need to subsist. To distinguish such situations from a kindred kind of dilemma that will be discussed under 'Desires in conflict', below, I shall call them cases of *needs in conflict*. Here is a hypothetical example:

> *The case of the poor society*. In the poor society, political authorities face the choice between two economic policies. They may either (i) open coal mines that will give work and income to destitute city dwellers, but destroy the natural habitat from which rural tribes derive their livelihood, or (ii) preserve the environmental basis of traditional ways of life at the cost of continued urban poverty. What is to be done?

Situations like this call for an extension of the previous argument. The priority principle provides no guidance on the hard choice facing the poor society. It rules out either option, but it has to be one of them.

If loss of subsistence will result anyway, it seems reasonable to do what leads to the least serious loss. This presupposes not only that we can discriminate, *ex post facto*, between the relative badness of different outcomes, but even that we can determine in advance whether one possible outcome will be better or worse than another. Thus, in the case of the poor society, we must be able to say whether or not the continued destitution of city dwellers will be worse than the plight that members of rural tribes will suffer if mining destroys their natural habitat.

Two things might justify the assessment that, among two states of affairs which both involve loss of subsistence, one is worse than the other: (i) the greater severity of the suffering taking place or (ii) the larger number of people afflicted by it. As far as (i) is concerned, discrimination may be grounded in the difference between death and nonfatal deprivation. I shall, however, set no store by this criterion when it comes to resolving cases of needs in conflict, as loss of subsistence is always liable to bring death in its wake. What, then, about (ii)? Can we resolve such hard cases by distinguishing between adverse outcomes according to the number of people whom loss of subsistence will afflict? The question is really twofold.

First, can we, in evaluating deadly policies, foretell the numbers of prospective victims with any precision? Although accurate predictions are mostly unavailable, rough estimates need not be. They may be based on estimates of the geographical extent of areas that are vulnerable to loss of subsistence. Provided we know enough to make such estimates, numbers can be predicted with some precision.

Second, do differences with respect to the number of prospective victims matter from a moral point of view? Can they serve the purpose envisaged here: ascertaining what to do when every available policy will leave someone without the means of meeting vital needs? If so, the following rule of action suggests itself: where every available policy will lead to loss of subsistence, take the one that has the smallest number of expected victims. This will be called the principle of *minimizing expected mortality*. It is, I believe, basically sound. Its rationale is brought out by an argument of F. M. Kamm's (1993: 85), called the *aggregation argument*:

> If (1) it is worse if B and C die than if B alone dies ...; and (2) it is equally bad if A alone dies or if B alone dies ...; then (3) by substitution, it should also be worse if B and C die than if A alone dies. That is, where '>' = 'worse than,' if B + C > B and if A = B, then B + C > A.

Proposition (1) seems self-evident. The death of two persons is worse than the death of only one of the two. Proposition (3) does not have the same appearance of self-evidence. The reason is presumably that the death of a single person is as bad as the death of two others from the single person's point of view. Granted this, one might question whether (3) is equivalent to (1) and, hence, whether there is always greater evil in loss of subsistence that befalls a greater number of people. A negative answer to this question would undermine the aggregation argument as well as the principle of minimizing expected mortality.

Such an answer has been advanced by John Taurek (1977). He dismisses the idea that numbers count in cases of needs in conflict. If a choice must be made between two actions, X and Y, and X will take away many people's means of survival while Y does the same to only one person, then the fact that more people will die if we do X is in itself no reason why we should do Y rather than X. In such trade-off situations, we should not 'consider the relative numbers of people involved as something in itself of significance in determining our course of action' (*ibid.*: 94). The reason lies in a distinction between two types of loss involved in someone's death. First, there is *loss to* the person who dies: he loses what is most valuable to him – his life. Second, there is also *loss of* a person, felt by those who remain to register and experience that the person in question is no longer around. The two types of loss derive from the same event, but there is an important difference between them: the loss that death brings *to* a person is apprehended by him only, and such losses are not additive. Thus, if P dies, the loss to her is as great as it can be; it would be no greater if she were to die in an accident that took many lives rather than suffer solitary death. Losses *of* people are, however, additive. The more deaths, the greater this type of loss.

So far, Taurek's argument is noncontroversial, but then he invokes a consequential and controversial premise: if every action available to us will lead to someone's death, our choice should be premised on no other considerations than losses *to* the persons whose lives are at stake. Those who take the opposite view, and (in line with the principle of minimizing expected mortality) would rather sacrifice one person than letting many die, are said by Taurek to attach 'importance to human beings and what happens to them in merely the way I would to objects which I valued' (*ibid.*: 306). To bring out this point, imagine six objects of equal value, five in this room and one in that room. All are threatened by fire, and we can retrieve only the contents of one room. What should be done? We should retrieve the five rather than the one, because they are five times more valuable, and their extinction would involve a fivefold greater loss of value. According to Taurek, however, this line of argument is improper if human beings are substituted for inanimate objects in the example.

> My concern for what happens to them is grounded in the realization that each of them is, as I would be in his place, terribly concerned about what happens to him. It is not my way to think of them as each having a certain *objective* value, determined however it is we determine the objective value of things, and then to make some estimate of the combined value of the five as against the one.
>
> (*Ibid.*: 306–7)

What, then, are we to do if we have to choose between an action that leads to one person's death and one leading to the deaths of many? In Taurek's view, we should show equal concern for everyone affected by the choice,[6] giving each person an equal chance of survival, which may be done, for example, by flipping a coin to decide if the one or the many are to be saved.

To defend the principle of minimizing expected mortality against Taurek's argument, we may start by insisting that losses *to* people who die are not the only thing that matters about their deaths; loss *of* life matters as well. Thus, John Sanders (1988: 12) avers that he 'cannot help feeling that the world is a better place with people in it than it would be without them.... People are valuable. They are worth saving because they are people. They *are* objects, although they are surely rather special objects. That is what makes them so valuable.' In Sanders' opinion, his disagreement with Taurek on this score boils down to a 'war of intuitions' (*ibid.*: 13). Both the notion that numbers count and its rejection reflect important strands of moral thinking: on the one hand, equal concern for everyone whose life is at stake, on the other hand, an acknowledgement that human lives are valuable. Sanders suggests, accordingly, that both points of view be represented in moral deliberation. His proposal is, more specifically, that numbers are a decisive consideration if and only if the choice is between actions whose respective tolls differ appreciably:

> if the choice is between saving one and saving five, one should flip a coin ... in *such* cases, consideration of losses-of is simply outweighed by [the] principle of equal concern. Nevertheless, in cases ... where the choice is

> between saving one life and saving *huge* numbers of lives, Taurek's view loses all intuitive plausibility.
>
> (*Ibid.*)

I believe Sanders is right about the counterintuitive character of Taurek's argument. The badness of death cannot be assessed only from the perspective of the dying. But is Sanders' informal compromise between Taurek's argument and the aggregation argument more plausible than the latter, which simply says that things are worse the more people die in exact proportion to the number of deaths? Consider four states of affairs: A involves one death; B involves X deaths, where $2 \leq X \leq 1{,}000$; C involves 1,000 deaths; and D involves 1,001 deaths. According to Sanders, C is worse than worse A, and B must likewise be deemed worse than A for some value of X. But Sanders does not deem D worse than C. Now, these states are alike except for one thing – the death of one person – which is conceived as immaterial when it comes to moral evaluation. Yet in another context – that of A and B – the same death might make a difference. It might be (part of) the reason why B is worse than A. I find this double perspective illogical and intuitively problematic. It is more plausible to say that one more death always makes things worse than they would have been in its absence. To be sure, if very many people are already dead, one more adds very little to the evil at hand, but this is not to say that it makes no difference at all. It makes all the difference of another fatality. Grisly as these matters are, one would rather not think about them, but if one has to, the principle of minimizing expected mortality seems the best place to start.

In concrete cases, however, we will often be incapable of confident assessments of whether and how much one state of affairs distinguishes itself from another in terms of expected mortality. Forecasts on this score are apt to be uncertain. I have, to be sure, hinted at one possible basis for estimating the number of people whose lives may be destroyed by a policy bringing loss of subsistence in its wake: the geographical extent of the territory affected by it. But if this defies reasonably precise demarcation, there is little to go on in applying the principle of minimizing expected mortality. As far as environmental destruction is concerned, I suspect that lack of precision on this score is the rule rather than the exception. This is to say that cases of needs in conflict often cannot be resolved by pursuing the policy that leads to the least serious loss of subsistence.

There is no way of telling which is better or worse among the outcomes of alternative policies.

How, then, should a choice be made? We might seize on Taurek's idea of showing equal concern for everyone affected by the choice. Thus, by flipping a coin to decide the matter, each person whose life is at stake will be given an equal chance of survival. Politically, this idea has no chance of being taken seriously today or in the foreseeable future, but from a moral point of view it seems the best way of resolving cases of needs in conflict that escape the principle of minimizing expected mortality.

So far, cases of needs in conflict have been portrayed as choices under certainty in which we know that every available policy leads to loss of subsistence. But what if we know only that such loss is one of many possible outcomes of every alternative before us? What, in other words, if a hard case has to be resolved under ignorance? Consider the following situation:

	States of the world					
	S_1	S_2	...	S_j	...	S_n
Acts	*Outcomes*					
A_1	O_{11}	O_{12}	...	O_{1j}	...	O_{1n}
A_2	O_{21}	O_{22}	...	O_{2j}	...	O_{2n}

If A_1 and A_2 are the only alternatives to choose between and at least one possible outcome of either action involves loss of subsistence, then the matrix above depicts a hard choice under ignorance. The rational way of dealing with it depends on what mode of ignorance we are talking about. This chapter has offered a tripartite typology: numerically estimable risk, first-degree uncertainty, and second-degree uncertainty. We must find out what rationally ought to be done with cases of needs in conflict under each mode of ignorance.[7]

In a situation of second-degree uncertainty, where nothing is known about the likelihoods of possible outcomes, we should choose the act whose worst outcome is least bad as judged by the principle of minimizing expected mortality. Making sure that there are no regrets if worst comes to worst seems reasonable when some outcomes would be particularly bad, as they would in a case of needs in conflict.

In decision theory, this approach is known as the *maximin principle*. It tells us to compare the minimum values associated with all alternatives and act with a view to maximizing the minimum value. Thus, 'each act is appraised by looking at the worst state for that act, and the "optimal choice" is the one with the best worst state' (Luce and Raiffa 1957: 278). Formally, let min O_{1j} denote the worst outcome that can result from A_1. The maximin principle says that A_1 is the preferred action if and only if min O_{1j} > min O_{2j}, for all j. If the two actions have worst outcomes that are equally bad, the natural thing to do is to compare them according to their next-worst outcomes, opting for the one whose next-worst outcome is best. If we cannot tell which is worse among their worst outcomes, the maximin principle does not single out one action in preference to the other, and we should, as pointed out earlier, resort to randomization.

The maximin principle is the only rule of rationality that can be applied with good reason to hard choices under second-degree uncertainty. It is applicable under first-degree uncertainty too, but such situations, in which the likelihoods of possible outcomes can be ranked on an ordinal scale, permit the use of two more decision criteria. First, knowledge of probabilities may supplement the maximin criterion as a basis of rational choice. If (i) A_1 and A_2 are equally good as judged by the maximin principle (their worst outcomes equally bad), and (ii) the worst outcome of A_1 is less likely than the worst outcome of A_2, then we should opt for A_1. The best worst outcome of this alternative is the least likely of its kind. This may be called *minimizing the probability of maximin*.

The second way of utilizing ordinal probabilities is apt to be more controversial. Imagine a case of mitigated first-degree uncertainty, in which we can not only rank the likelihoods of possible outcomes, but even say something about the difference between them (see above). Assume that O_{11} and O_{22} are the worst outcomes of A_1 and A_2, respectively. O_{22} is somewhat worse than O_{11}, but not very much. On the whole, it makes little difference which of these results comes about. Assume, moreover, that the likelihood that O_{22} obtains if A_2 is chosen is markedly lower than the likelihood that O_{11} comes about if we opt for A_1. Thus, choosing A_2 yields an appreciably greater probability that something better than either O_{11} or O_{22} takes place. If so, it seems rational to choose A_2, although A_1 is the maximin solution. By running the risk of a bad outcome which is slightly worse than the best worst outcome, we minimize the likelihood of any outcome rated so low.

This rule of rational action was introduced by Gregory Kavka (1987) under the name of the *disaster avoidance principle*. It applies, in Kavka's view, as long as the worst possible outcomes of two courses of action are both 'regarded as extremely unacceptable, that is, as involving very large amounts of negative utility'. Then the 'truly disastrous nature of the lesser disaster makes it sensible to risk the worse disaster in the hope of avoiding all disasters. The wisdom of following MMP [the maximin principle] is quite doubtful under these conditions, as this would maximize the probability of there being an extremely unacceptable outcome' (Kavka 1987: 67–8; italics omitted).[8]

What, finally, if a hard choice is to be made under numerically estimable risk? Then numerical estimates can be assigned to the possible outcomes of alternative policies and the *expected utility principle* recommends itself as a criterion of rationality. The expected utility of an action is found by first multiplying the value (or utility) of every possible outcome of this action with the likelihood that it comes about and then summing the products. To maximize expected utility is to choose the act, A_i, that maximizes $\Sigma p_j U(O_{ij})$, for all $j = 1 \ldots m$, where p_j is the probability that O_{ij} results from the choice of A_i, U is a utility function defined over possible outcomes, and $U(O_{ij})$ represents the value for O_{ij}.

The expected utility principle can be used only where the values of possible outcomes are measurable on an interval scale. Evaluations must, in other words, be capable of representation by a cardinal utility function. This function would have to reflect the priority principle and the principle of minimizing expected mortality, and these can be translated into a cardinal utility function by defining utility as a negative, linear function of the number of expected deaths. This squares well with the earlier elaboration of the principle of minimizing mortality. As pointed out, however, the extent of expected mortality resulting from loss of subsistence often defies prediction. If it does, the utility function is undefinable and the expected utility principle finds no application. We must fall back on the maximin criterion or the disaster avoidance principle.[9]

Desires in conflict

If there is no risk that anyone will be deprived of subsistence whichever way an environmental risk is handled, the priority principle provides no guidance and we face another kind of hard case. Here is an imaginary example of a situation in which vital needs will be met all round, and only desires are at stake.

> *The case of the affluent society.* In the affluent society, political authorities face the choice between (i) improving the road system by building new highways through forests and woodlands and across mountain terrain, and (ii) foregoing such improvements in order to save remnants of pristine nature. What is to be done?

This may be called a case of desires in conflict. There are basically two ways of answering the question it poses. The first is to find out what people of the affluent society desire – whether they want a better road system to make transport faster and cheaper, or prefer preserving areas of wilderness – and then aggregate individual desires to arrive at a judgement as to what is best overall. In the philosophical literature, *utilitarianism* offers by far the most famous method for doing this. It consists of aggregating desires on the basis of how widely and strongly they are felt. Individual levels of welfare or utility are added up to find out whether one state of affairs – say, the one that comes about if the affluent society opts for alternative (i) – is better or worse than another – say, the one resulting from the adoption of alternative (ii). One state is deemed better than another if its aggregate welfare is greater, and we should act so as to maximize aggregate welfare.

Ordinary renderings of utilitarianism presuppose interpersonally comparable, interval-level estimates of how well people fare. The technical obstacles to obtaining such estimates are daunting, and utilitarianism is often rejected on epistemological grounds. This problem need not concern us here, however. Utilitarianism is primarily invoked to illustrate one way the priority principle might be supplemented with a principle applying to environmental dilemmas involving desires in conflict. (True utilitarians, of course, apply the method of sum-ranking wholesale to conflicts of interest, making no distinction between interests of different normative force.) The crucial thing is not to distinguish between desires according to contents. Everything is converted into a common currency – social welfare. Thus, in the case of the affluent society, the inherent characteristics of policy alternatives are irrelevant to the choice between them; all that matters is the aggregate level of welfare attainable by pursuing either (i) or (ii).

Among existing political arrangements, the closest parallel to choice on the basis of the utilitarian principle is democratic decisions based on majority voting and universal suffrage. The only difference – which, of course, is all-important in some respects – is that the democratic procedure aggregates individual preferences without taking their intensity into account. For our purpose, the affinity between this procedure and the utilitarian method of aggregation lies in the fact that neither treats any desire as more worthy of fulfilment than another. Both take what we may call a *pluralist* view of preferences.

If we depart from the neutral view, a second way of answering the question posed by the case of the affluent society suggests itself. Suppose that the fulfilment of a certain desire – say, the desire that pristine nature be preserved intact – is pronounced more important than the fulfilment of others, including the want of ever greater affluence. Add to this the assumption that policies catering to the worthiest desires are preferable to policies providing for the less worthy ones. Then you have a *perfectionist* principle of political choice. It distinguishes between desires according to contents, but takes no account of how widely or strongly various policies are wanted. Certain goods are put before others in terms of importance regardless of what people think about them.

A perfectionist supplement to the priority principle can take many forms, depending on how desires are being graded. The principle hinted at in the previous paragraph implies that the preservation of pristine nature is preferable to its further utilization when vital needs have already been met. Is this a plausible point of view? And is perfectionism in general a valid approach to public policy? To take the last question first, perfectionist ideas seem least controversial when they echo most people's considered, if not immediate, judgements. Thus, Adam Smith suggests that certain physical strains are truly worse than emotional ones, although the latter are more widely and strongly lamented.

> The loss of a leg may generally be regarded as a more real calamity than the loss of a mistress. It would be a ridiculous tragedy, however, of which the catastrophe was to turn upon a loss of that kind. A misfortune of the other kind, how frivolous soever it may appear to be, has given occasion to many a fine one.
>
> (Smith 1759/1976: 29)

Reverting to environmental matters, it is by no means obvious that the loss of pristine nature is a more real calamity than an improvement foregone in the material standard of living. When the experience of 'a nature that is raw, wild, untainted by man' – as Daniel J. Kevles (1989: 32) puts it – is juxtaposed to the prospect of harnessing natural processes, it may simply seem that two different and essentially contradictory forms of life stand before us. One is espoused by those who want to preserve unspoiled nature so that they, in the words of Bill McKibben, 'can worry about ... human affairs secure in ... knowledge of the eternal inhuman' (quoted in Kevles, *ibid.*). The other is highlighted in the praise that the *Economist* sang to Los Angeles after a major earthquake hit the city in January 1994: 'Here is a city that sits on a comprehensive system of geological faults; yet, it goes on growing, spinning off great concrete webs of suburbs and freeways, as if it were perfectly safe.... No other city has challenged nature so persistently – and, in general, so successfully' (22–28 January 1994: 14). Rather than resenting the taming and transformation of wilderness for human purposes, this technological feat is, according to the *Economist*, 'precisely the point of Los Angeles': it is 'an affront to nature ... [a]nd all the better for it' (*ibid.*).

Taking a pluralist view of preferences, these outlooks will simply present themselves as opposed points of view that merit equal consideration. If so, we must decide between them by way of aggregation, and while there will inevitably be disagreement about the proper method for doing so, there is no need to raise fundamental questions about the intrinsic superiority or inferiority of different conceptions of good and bad. It may be objected that pluralism has already gone by the board in this book, which ranks vital needs over desires in terms of normative force. However, this ranking rests on an argument that, like Smith's, purports to distinguish real calamities from spurious ones in a way which sits well with widely divergent conceptions of what makes life good or meaningful. The theory of interest does not run counter to moral pluralism in the way of theories that see inherent value in either the 'eternal inhuman' or technological accomplishments. It appeals to an underlying consensus.

Pluralist qualms notwithstanding, I believe there is a basis for favouring one of the views that stand opposed in the case of the affluent society. The want of preserving nature has greater merit than the desire for further improvements in material welfare, not on the ground that the former is an intrinsically superior conception of good and bad, but because affluent societies have generally gone so far in accommodating the quest for harnessing nature that they are about to eradicate the opportunity of experiencing wilderness. It is the risk that further economic growth may eliminate a certain form of life as a live option which makes preservation so important. Thus, regardless of how widely and strongly the desire for further utilization of nature is felt, it must be turned down before it is too late. There is no metaphysical reason for doing so; the rationale lies in historical facts about a shrinking matrix of eligible ways of life. This line of argument departs from the pluralist view of preferences only in saying that the continued existence of the *opportunity* of experiencing pristine nature counts for more than the actual fulfilment of material desires.

The importance of preserving pristine nature in contemporary affluent societies may be justified in basically the same way that the urgency of protecting of civil rights is defended. According to liberal political theory, the protection of such rights comes before other concerns, notably the accommodation of the majority view in democratic decisions. Constitutional constraints are recommended as barriers to policies that, although highly popular at the moment, will later be regretted but cannot then be undone. Thus, the 'temporary suspension of rights easily leads to the permanent abolition of majority rule itself and its replacement by dictatorship. It suffices to cite the years 1794 and 1933' (Elster 1988: 9). I believe constraints ought likewise to be placed on a democratic system to ensure that the pursuit of technocratic goals does not place the opportunity of experiencing untamed nature beyond reach. The taming and transformation of nature must not be allowed to vitiate people's freedom of choice by leaving them with a considerably truncated matrix of opportunities.

What practical implications does this argument have with respect to environmental policy in affluent societies? It hardly justifies an absolute ban on further technological inroads into the natural environment. The constitutional analogy rather suggests that procedural limits should be put in place to slow down the decision-making process when issues of the kind considered in this section present themselves. An affluent society should not allow itself further material advantages at the expense of pristine nature with the same ease that it otherwise passes political decisions. It should guard against further erosion of the opportunity of experiencing wilderness.

The modalities of the procedural limits remain to be sorted out, but the general idea is clear: where vital needs will be met regardless of how an environmental dilemma is resolved, the decision not to avert risks of environmental destruction should be hedged with restrictions corresponding to constitutional constraints on the suspension of civil rights.

Conclusion

This chapter and chapter 2 present a normative theory which shows the relevance of interest-based considerations to the resolution of environmental dilemmas. The general idea is that those whose interests suffer the greatest setback should be the subject of strongest concern. This implies, in particular, that vital needs should be safeguarded in a choice between policies that may either lead to unmet needs or unfulfilled desires, provided the risk that needs will go unmet is significant. This is the priority principle. It has been supplemented above with two normative standards pertaining to hard cases for which this principle is inapplicable. The first relates to cases of needs in conflict and says, in essence, that if vital needs may go unmet anyway, the fewer the better. The second was elaborated to deal with dilemmas in which only desires are at stake. In such cases, the preservation of pristine nature has a particular importance that should be reflected in the way environmental policy is made.

Notes

1 This is confirmed by the fact that American policy makers made divergent assessments of the chance that military intervention in Cuba would trigger nuclear war. President Kennedy saw a high risk, while most members of his government were less alarmist and warned against an 'exaggerated concern' which prompted 'hesitations where none were necessary' (according to an internal government report on the crisis quoted by Allison 1971: 218).

2 Another argument associated with Bayesian decision theory should, however, be categorically rejected. It goes as follows. If we are totally ignorant about the likelihoods of different states of the world, it is, for all we know, no more likely that one state obtains than any other. Each state must therefore be deemed equally likely. This is the *principle of insufficient reason*. Applying it to a situation in which the world may be in m possible states, it assigns a probability of $1/m$ to each state. This line of argument leads to contradictions if there is no natural division of states, as is mostly the case. Speculating about the political future of Russia, I may take the set of possible states to be {a military coup takes place; no military coup takes place}, which implies that the probability of a coup is 0.5. Or I may start from the set {a successful military coup takes place; an unsuccessful military coup takes place; no military coup takes place} – which is not another set of possibilities, but the same set described differently – implying that the chance of a coup is 0.66. I will, in other words, assign different probabilities to the same state of affairs depending on how the set of possible states is described. This is why the principle of insufficient reason must be rejected.

3 But knowing what I do about p_1, p_2 and p_3 is not compatible with knowing nothing about p_3 and p_4 except that p_4 is greater than p_3. Completeness implies that I also rank p_4 in relation to p_1 and p_2, and having done that I will be able to say something about the difference between p_3 and p_4.

4 Lennart Sjöberg (1979: 53) remarks that in all technological systems, there is 'a residual of unanticipated events brought about by human mistakes or sabotage or by complex and unanticipated interdependencies that may bring about disaster'.

5 When the parliamentary opposition in England met in 1647, after Charles I had been taken prisoner and the Civil War came to a critical juncture, the 'future looked remarkably wide open, the context of English politics extraordinarily fluid'. Don Herzog (1989: 7) continues: 'Had one of the age's many prophets managed to reveal the actual future, one and all would have been surprised. Charles would escape in November, and war would break out again; in 1649, after decisive Parliamentary victories, he would be put on trial and executed; the House of Commons would take charge, abolishing monarchy and Lords, and would attempt to transform England into a Puritan commonwealth; that attempt would collapse, and the army would come close to assuming control; finally the failure of all efforts to hammer out a stable regime would lead almost magically to the restoration of Charles II in 1660.'

6 Taurek adds that emotional ties or contractual obligations may give ground to favour someone in particular. He allows the principle of equal concern to be overruled by some seemingly lightweight considerations, like the fact that we happen to know some of the persons involved. I shall disregard this part of his argument, which mostly proceeds by way of unargued assertions. But it touches on a subject that will be taken up in the next chapter – the normative force of noninterest-based considerations rooted in special relationships between people.

7 People act rationally to the extent that (i) their goals are consistent, (ii) their beliefs about how these goals can be attained are grounded in comprehensive and reliable information about available alternatives, and (iii) they opt for the alternative they believe to be best. If they know for certain what the effect of any given alternative will be, there is no question as to what rationally ought to be done, but if they stand before a decision under ignorance, things are less straightforward. I shall recommend some decision criteria, and one may ask whence the general notion of rationality that

underlying these comes? Jon Elster (1989: 3) argues that 'the notion of rationality has to be independently plausible as a normative account of human behaviour,' and 'it has to yield prescriptions about particular cases that fit our preanalytical notions about what is rational in such cases'. This is to say that a list of decision criteria should rest on (i) general beliefs about reasonable and unreasonable attitudes to ignorance, and (ii) particular beliefs about the wisdom or unwisdom of displaying specific attitudes in concrete situations. Is there much overlap in the assumptions people make on these scores? Do they think alike about rationality in general and the rational handling of particular problems of ignorance? To some extent they do, but the overlap is by no means perfect, and there are no proofs we can turn to in order to settle remaining disagreements about rationality and irrationality. I only hope that the argument below accommodates most reasonable attitudes to ignorance.

8 Kavka invokes one further condition for applying the disaster avoidance principle. It comes into the picture only if we are capable of not just ranking and comparing probabilities, but also to some extent estimating them in absolute terms. If so, we should apply the disaster avoidance principle only if the probabilities of the two relevant adversities are not negligible. If they are negligible, it seems unwise to be exclusively preoccupied with adverse outcomes as opposed to more desirable ones that are much more likely to come about. This judgement may be made only in case of mitigated first-order uncertainty that comes close to numerically estimable risk. It then seems reasonable enough, as there is little sense in preoccupying oneself with extremely unlikely disasters.

9 It will be noted that hard choices under numerically estimable risk are extremely rare, granted the dismissal of subjective probabilities in the section 'Numerically estimable risk'.

4

Realism and responsibility

Introduction

If the urgency of interests is all that matters, the correct way of resolving environmental dilemmas depends exclusively on the priority principle and interest-based standards pertaining to hard cases which this principle cannot resolve. I shall argue, however, that other standards must also be taken into account before overall conclusions on what to do in concrete cases can be reached. While the theory of interest has so far ruled alone, this chapter brings in countervailing considerations.

Two aspects of the argument of chapters 2 and 3 restrict its purview, as it were. In the first place, normative problems are approached from a purely *impersonal* perspective. The urgency of meeting various human interests is assessed without consideration of whose interests we are talking about. According to this view, there is no way I ought to care more for someone than the objective force of her interest warrants because I stand in a special relationship to her. Whether she is my fellow Norwegian, my mother, my friend or myself, I should not accord greater weight to her needs or desires than to anyone else's. This chapter will challenge this perspective, inquiring whether the impersonal concern about human interests has to be supplemented with noninterest-based considerations pertaining to personal relationships. At the extreme, such considerations might imply that the vital needs of strangers can be disregarded whenever they can be met only by dint of sacrifices to ourselves or our nearest and dearest. According to a more moderate version of the objection, an undifferentiated concern about humanity is unrealistic and untenable. This is, I believe, a valid point of view, but its precise implications for normative assessment are hard to tell.

In the second place, the argument of chapters 2 and 3 is forward-looking, in that it sets store only by the subsequent effects of activities and policies. Thus, there is no question of whether or not a given action is an appropriate reaction to events that lie in the past. Sometimes, however, historical information is relevant when it comes to telling right from wrong. If, for example, some people are responsible for the very existence of an environmental dilemma – if, that is, previous actions of theirs have landed them and others in such a situation – they have a special responsibility for shouldering what it takes to get out of it. So, at any rate, I shall argue at the end of this chapter.

Special relationships and realism

The theory of interest represents an impersonal outlook on moral problems. It defines the normative force of needs and desires without regard to whom they belong. One may object that important considerations escape such an outlook. This is not to say, of course, that the theory of interest is inadequate in itself; there is no questioning the central place of vital needs in moral thinking or the role of the priority principle as a guideline of environmental policy. But the theory of interest does not necessarily exhaust the moral point of view.

Lawrence Blum (1988: 473) writes that 'each person is embedded within a web of ongoing relationships, and morality importantly if not exclusively consists in attention to, understanding of, and emotional responsiveness toward the individuals with whom one stands in these relationships'. The purely impersonal outlook is, more specifically, vulnerable to what has been called 'the nearest-and-dearest objection':

> Our lives are given shape, meaning and value by what we hold dear, by those persons and life projects to which we are especially committed. This implies that when we act we must give a special place to those persons (typically our family and friends) and those projects.
>
> (Jackson 1991: 461)

Granted the nearest-and-dearest objection, it follows that people may reasonably accord greater weight to their own interests and the interests of those who have a special place in their lives than they would have done if their only concern were the normative force that interests can be accorded from an impersonal point of view. This will be called the principle of *reasonable partiality*. For simplicity, I shall hereafter speak of all source of reasonable partiality as *personal relationships*, including, *inter alia*, kinship, friendship and the relationship that binds a person to his or her own projects.

A notable implication of the principle of reasonable partiality is that the vital needs of strangers may count for less than the desires of relatives, friends, compatriots or oneself. To be sure, nothing I said above proves that the principle ever goes this far. Impersonal considerations definitely do not disappear from the picture the moment personal concerns enter into it, and conclusive judgements about right and wrong will reflect a blend of the two perspectives. But the possibility of overturning moral priorities that can be established on an impersonal basis arises. Albert Camus puts it tersely: 'I believe in justice, but I will defend my mother before justice.'[1]

What, then, can be said for the principle of reasonable partiality? Is the nearest-and-dearest objection valid? To some – the most famous of whom is William Godwin (1798/1976: 170) – partiality definitely does not belong in moral deliberation:

> What magic is it the pronoun 'my', that should justify us in overturning the decisions of impartial truth? My brother or my father may be a fool or a profligate, malicious, lying or dishonest. If they be, of what consequence is it that they are mine?

One notable consequence is that I may be humanly incapable of assessing their situation and judging their behaviour objectively. Much as I appreciate the verdicts of the impersonal point of view, sympathy for relatives and friends makes me biased in their favour and my commitment to personal projects renders me reluctant to give them up on grounds of regard for strangers. People are, to be sure, differently disposed on this score, as the contrast between Camus and Godwin illustrates, but one would be hard put to dispute that 'the common and natural course of our passions' – as David Hume (1740/1985: 483–4) calls it – blunts the capacity for caring about people only in proportion to the impersonal force of their interests. Some odd characters notwithstanding, there is no denying that a 'man naturally loves his children better than his nephews, his nephews better than his cousins, his cousins better than strangers, where everything else is equal' (*ibid.*). This is armchair psychology, but however much scientific studies of motivation might refine the diagnosis, they will hardly overturn the observation that, as long as the stakes are nontrivial, very few people can be expected to put their own interests or the interests of relatives, friends and compatriots on a par with those of strangers.

Briefly put, the argument just indicated says that normative assessment must be premised not just on impersonal considerations but also on the principle of reasonable partiality if it is to be done in a spirit of realism. Were the impersonal perspective to take over entirely, judgements of right and wrong would be apt to lose contact with reality by demanding too much in terms of identification with the plight of strangers. I shall call this the *realist* argument.

Another argument for the inadequacy of purely impersonal assessments is this: people ought to care more about the interests of their nearest and dearest than an objective evaluation warrants because special devotion to personal relationships is valuable. Thus, it is not only unrealistic to expect that people will show the same concern for strangers that they show for themselves and their families, they would be wrong (or at least blameworthy) to do so. We may call this the argument that *partiality is positive*. Its most eloquent defence is mustered by Bernard Williams (1975: 116), who asks: how can a person 'come to regard as one satisfaction among others, and a dispensable one, a project around which he has built his life'? To demand that he 'just step aside from his own project and decision' when impersonal considerations call for it, would be 'to alienate him in a real sense from his actions and the source of his action in his own convictions'; it would be 'an attack on his integrity' (*ibid.*: 116–17).

I shall pass over the argument that partiality is positive for two reasons. First, it does not, at least in Williams' statement, go very far in lending personal interests greater weight than they can be seen to have from a purely impersonal point of view. After all, only a project around which someone has built a life is thought be one whose renunciation involves alienation and loss of integrity. But many demands that cannot reasonably be charged to this account will fall on deaf ears because people are naturally biased towards their nearest and dearest. Thus, the realist argument, if valid, will make the principle of reasonable partiality a more substantial counterweight to impersonal considerations than the argument that partiality is positive can do.

Second, and more importantly, the argument that partiality is positive can hardly be justified without crucial support from the realist argument. Consider Williams' claim that acquiescence to the demands that derive from impersonal considerations may alienate a person from her or his actions and their sources in convictions. This would not be the case if the common and natural course of our passions was, *pace* Hume, not to care particularly for our nearest and dearest, but rather to regard personal projects and relationships as ephemeral phenomena: things we become attached to for a while but easily forget about. Then, of course, a person would not build a life around a single project or relationship; life would rather evolve like a conversation that moves constantly from subject to subject. And then the demands that derive from impersonal considerations would not threaten anyone with alienation, as no one would ever be in a position to give up 'projects and attitudes with which he is ... closely identified' (*ibid.*).

As things are, people do identify closely with particular projects and persons, and impersonal considerations are a potential threat to such commitments. It is, accordingly, the way the world is – brute facts about human motivation – which ultimately lie behind the argument that partiality is positive. It is positive because it is a prerequisite of living a normal human life constituted by enduring personal projects and lasting relationships. If another kind of life had been a realistic alternative for most people, the value of living like we actually do would be an open question. And if, as suggested in the last paragraph, a life of lasting commitments had been abnormal, few if any would find it valuable.

To conclude, there is something to the argument that partiality is positive if and only if there is something to the argument that normative assessment should be imbued with a spirit of realism. It is valuable that normative assessments somehow respect the normal and natural course of our passions because this course inevitably constrains the way we think about normative problems and must be respected in order to forestall alienation. Were the nature of human beings not (in part) defined by a specific course of their passions, arguments about alienation would be quite inapposite. It is possible, to be sure, that the argument that partiality is positive could be restated so that it no longer needed crucial support from the realist argument, but I do not know how this should be done and will therefore stick to the realist argument when it comes to justifying the principle of reasonable partiality.

Is there anything to the realist argument? Should normative assessment be done in a spirit of realism? Why not rather do it in the idealist spirit, summoning moral theory as a radical corrective of human passions?

I know of no argument that proves either realism or idealism to be the proper spirit of moral thinking. But realism may be defended indirectly: it underlies an attractive conception of the place of moral ideas in human life. This is spelled out by Samuel Scheffler (1992), who contrasts it to the conception underlying purely impersonal principles. The latter answers to the idea that 'morality represents a form of self-transcendence', the former to 'the idea that morality makes possible an important form of personal integration' (*ibid.*: 124). Only the former implies that living morally 'is motivationally accessible to normal moral agents: ... a serious if not always easy option for normally constituted agents under reasonably favorable conditions' (*ibid.*: 125). The attractiveness of realism lies in the fact that it dispels a potentially glaring rift between moral principles and human motivation. Idealism, by contrast, may lend an air of practical irrelevance to moral thinking. But, it may be countered, what are moral demands if not earnest requests to reconsider conventional and convenient ways? If people resist radical appeals, is it not all the more reason to insist that they be taken seriously? Such insistence will at least disturb people's peace of mind and may gradually incite some movement towards reconsideration and personal reform. Sudden illumination is conceivable, as Stuart Hampshire (1989: 103) points out, taking Paul on the road to Damascus as his model. But there is also a risk that radical appeals will be highly counter-productive. There is not just the straightforward problem that an exacting normative principle may be ignored and its recommendations prove irrelevant to deliberations and actions. If so, insisting on the principle is idle talk, but does no harm. What comes in addition is a risk of utopianism, which was first pointed out by Thomas Aquinas. Alluding to a passage in St Matthew, he argued:

> 'if they pour new wine' – that is, prescriptions of perfect behaviour – 'into old wineskins' – that is, imperfect persons – 'then the wineskins will break, and the wine is spilled' – which is to say that the prescriptions will fall into disrepute, and as a result thereof people will degenerate to become even worse than they were.
>
> (Aquinas 1967 edition, question 96: 132)

Aquinas was thinking of proposals for legal reform when he wrote this, but the argument carries over to morality. If we, as John Rawls (1971: 246) proposes, start with a blueprint of ideal arrangements and relegate the 'adjustments to natural limitations and historical contingencies' to some future stage of practical implementation, then it is possible that our ideals are taken seriously only by unprejudiced people of exceptional generosity. They may become what J. L. Mackie (1977) calls an 'ethics of fantasy'. He warns that identifying morality with 'something that certainly will not be followed is a sure way of bringing it into contempt – practical contempt, which combines all too readily with theoretical respect' (*ibid.*: 132). What looks like moral progress in abstract argument may, in other words, cause moral decay in social relationships. An exacting moral demand may give a bad name to morality as a whole, lending its injunctions an air of unworldliness that evokes a general attitude of moral scepticism.

This is no knock-down justification of the realist argument, but the risk of utopianism lends it considerable support – enough, I think, to warrant invoking the principle of reasonable partiality alongside the priority principle and other interest-based standards as a normative guideline for resolving environmental dilemmas.

Defining reasonable partiality

How, then, are we to incorporate the concern about personal relationships in normative assessment? Should it be merged with the theory of interest to make up a bifurcate – partly impersonal, partly personal – theory? I shall not take this course. The theory of interest and the realist argument are best seen as distinct perspectives on right and wrong, incapable of coalescence. I shall rather suggest a somewhat mechanical way of combining the priority principle and other impersonal standards with the principle of reasonable partiality to work out conclusive solutions to concrete problems.

Confronted with an environmental dilemma, the first question to ask is: what should have been done if impersonal considerations were the only ones that mattered? Then a reality check must be administered: how far may we reasonably go in demanding that people do what the answer to the first question says? Thus, in order to find a way out of an environmental dilemma, like the dilemma relating to climate policy, one begins by taking an objective view of all interests involved and proceeds to formulating realistic proposals for public policy. There is no saying, of course, that realism always dictates deviation from the course that recommends itself on impersonal grounds, but it may well do so. The crucial consideration on this score is that the things people are enjoined to do are feasible options for normally constituted agents.

How, then, should feasibility be assessed? How do we determine more precisely the extent to which we can permit ourselves to take a partial view of how the interests of our nearest and dearest compare with those of strangers? It can be done in basically two ways: by applying a general partiality test to concrete problems, or by evaluating each problem on its own merits. An example of the first approach is this: multiply the stake you have in an issue by a factor, *M*, which quantifies the extra weight people are generally allowed to accord their own interests, and do what is best for yourself only if your own stake multiplied by *M* exceeds whatever stakes other people have in the same issue.

> Imagine that I want to perform some act, S, rather than an alternative, O, because S is more in my interests.... performing S rather than O would be permissible even in cases where the loss to others does outweigh the gain to me, provided that the size of the loss to others is less than or equal to M times the gain to me.
>
> (Kagan 1984: 250–1; italics omitted)[2]

The problem is, of course, that this formula cannot be more than a quasi-precise restatement of the argument which went before. There is no way of actually quantifying *M*, and nor are losses and gains normally measurable in quantitative terms. In general, I strongly suspect that any attempt to translate the principle of reasonable partiality into a universal formula for administering the reality test will fail for basically the same reason: the circumstances surrounding concrete problems are too varied.

If so, the other approach – evaluating the realism of impersonal demands from case to case – may seem more promising. But even such evaluations will have to draw explicit or implicit justification from general ideas as to where the line between reasonable and unreasonable partiality goes. One idea that is often invoked is the notion that no one should be enjoined to do things that it would be virtually *impossible* for them to do. It is the weakest possible interpretation of the distinction between options that are feasible or unfeasible options for normally constituted agents. Hence, it is a natural place to start.

First, a moral demand must be deemed unrealistic if acceding to it is literally impossible. The tenet that 'ought implies can' provides a test all normative standards must pass, and the laws of physics point to a straightforward line between things that can and cannot be done.[3] But this does not take us very far. Any application of the priority principle, for example, falls on the safe side of the line. Second, however, there may be laws of psychology that provide less trivial criteria of what can and cannot be done. It is arguable, in particular, that an action is impossible if it is excessively costly or risky, even if its performance is not unfeasible in the literal sense. Felix Oppenheim (1987: 372) adduces the case of 'a bank teller [who], faced with a credible gun threat, cannot, for all practical purposes, refuse to hand over the money'. This proves that it makes good sense to speak of actions that are *practically impossible*.

Practical possibility is not a well defined concept that can readily be employed to weed out unrealistic moral demands. When 'the opportunity costs become so high or the risks so great as to render a contemplated course of action practically impossible or necessary' (*ibid.*) depends on the stakes involved as well as differences with respect to the risk that what is at stake will be lost by complying with a particular demand. In the case of a bank teller faced with a credible gun threat, life is at stake, and one definitely cannot demand of people that they do not succumb to armed robbery. We are, in effect, talking about a form of practical impossibility that borders on physical impossibility. When stakes or risk are not this high, there are no simple answers to questions of feasibility.

Oppenheim (*ibid.*) suggests that what he calls people's *basic preferences* are a determinant of practical possibility. We may assume that these preferences distinguish themselves from preferences *tout court* by being less malleable, turning on things that people have come to take for granted. To get an idea of their nature, recall the scholars described in chapter 2, who strongly desire an ample private library. While they suffer only inconveniences by being without it, one will be hard put to persuade them to divert their expenditures for literature to other causes. In general, the citizens of affluent countries have become used to many goods that few will abandon unless coercive measures bar their enjoyment – a varied diet, fast means of communication, long holidays, and so on. It may be argued that the renunciation of desires for such goods, while not practically impossible in a strong sense, is utterly unrealistic.[4] If so, the rejection of impersonally grounded demands for their renunciation is within the limits of reasonable partiality.

On this point, however, one should be wary of conceding too much to realism.[5] Basic preferences change, sometimes radically. A notable example, resulting partly from internal soul searching, partly from external influence, is the political transformation of Japan and Germany after 1945. Two countries bent on expansion were gradually recast to pacifist societies, which, according to Robert Jervis (1988: 343), shows that the 'most far-reaching changes in international politics involve changes in national goals and values'.

> Japan is now a much more suitable partner for cooperation than it was in the 1930s, and not only because territorial expansion is neither possible nor economically necessary. Something has occurred that is more basic than changes in instrumental beliefs. Rabid nationalism and the drive to dominate have been transmuted. A Japanese nationalist of the 1930s who saw his country today would be horrified, as Mishima was.
>
> (*Ibid.*)

In view of cases like this, what look like limits of practical possibility may rather reflect the 'limited power of the imagination to envisage an improved way of life with different commitments and purposes, which, once concretely envisaged, might be judged to be altogether better' (Hampshire 1989: 104). If so, an important purpose of normative thinking will be to aid the imagination in conjuring up radical visions.

Another observation casts additional doubt on the fixed nature of basic preferences. In Jervis' account of Japan's recent history, the transformation of national objectives is juxtaposed to changes in *beliefs* about factual matters. Jervis may be right that the latter are less basic than the former, but a change of belief can precipitate an alteration of preferences. Thus, one may cease to relish a dish after learning its composition. This mechanism can have far-reaching consequences. To take but one example, the unprecedented toll of World War I in terms human of life and material destruction altered people's conception of warfare. A. A. Milne wrote in 1935 that 'in 1913, with a few exceptions we all thought war was a natural and fine thing to happen.... Now, with a few exceptions, we have lost our illusions; we are agreed that war is neither natural nor fine, and that the victor suffers from it equally with the vanquished' (quoted in Mueller 1988: 75). This shows that an effort to change people's basic preferences may well take the indirect route of influencing their beliefs about what going after the goals in question really implies. Although people do not always believe what they ought to do in the light of available information, they are rarely incapable of reconsidering established opinions. The cognitive component of preferences can be a focus of moral exhortation.

But it is also possible that desires influence beliefs in ways which may impede efforts to influence basic preferences by working on their cognitive components. Thus, the seemingly innocent phenomenon of wishful thinking – desiring something so strongly that one is led to think there are ways of getting it although available information indicates that there are none – can have tremendous political implications. World War I once more offers an illustration. Before the war, political and military leaders in all the major European powers grossly overrated the operational and logistical advantages of offensive warfare, despite considerable evidence of its difficulties. Their optimism seems largely to have been fostered by urgent interests in changing the strategic status quo (Snyder 1984, ch. 1). The lesson to draw from this dramatic example of irrational belief formation is that one cannot always rely on the power of the better argument in efforts to alter people's beliefs. They may not, in particular, heed arguments whose truth would place their objectives beyond reach and call for revisions of their desires.[6]

All in all, no clear line can be drawn between options that are feasible or unfeasible for normally constituted agents once we go beyond the strict test of practical impossibility. The discussion of basic preferences provides only pointers to how far people can be expected to accede to – or at least take into account – demands deriving from purely impersonal judgements. This is to say that the principle of reasonable partiality cannot be stated with the level of precision that proved possible with respect to normative standards rooted in interest-based considerations. But it is precise enough to serve as a counterweight to these standards in overall assessments of how environmental dilemmas ought to be resolved. While specific implications of reasonable partiality must be worked out case by case, it will be done in the light of some general factors alluded to above: on the one hand, the risk of counterproductive utopianism; on the other hand, the prospect of moulding basic preferences – directly, through appeals for changes of basic preferences, or indirectly, by influencing people's beliefs about what the realization of their preferences actually implies.

Responsibility

The normative standards invoked in chapters 2 and 3 are forward-looking. They tell right from wrong on account of how alternative policies influence people's prospect of meeting needs and satisfying desires. The priority principle implies, in particular, that it will be wrong to pursue a policy that takes away someone's means of meeting vital needs if this can be avoided by pursuing another policy which only frustrates someone's desires. But what if people whose needs are at stake are themselves responsible for the existence of an environmental dilemma? Consider the following hypothetical example.

> *The federation's dilemma.* In a federation, F, all member states derive some of their export earnings from fisheries. Yet, one state, situated at the coast, is far more dependent on fisheries than the others. To the latter, they provide a welcome but not essential source of income; for the former, they are all-important. The federal government allocates fishery quotas that reflect the unequal situation of states in this respect. For several years, however, the coast state engages in overfishing to boost its export earnings. There is virtually no opposition to this policy among its inhabitants, although other states repeatedly warn about the risk involved. After some time, fishery resources are on the verge of depletion and the total annual catch must be cut down for a long period to allow regeneration. The federation faces a choice between basically two alternatives: either (i) to reduce each state's quota for several years, or (ii) temporarily to close down fisheries altogether in all states save the coast state, which is allowed a substantial annual quota. Alternative (i) implies severe economic problems for the people of the coast state. They avoid such problems if alternative (ii) is chosen, but then the other states of F will be adversely affected, although not nearly as seriously as the coast state in the case of (i). What should be done?

If it had not been for the folly of the people of the coast state, the federation's dilemma would not exist. There is no denying the relevance of this fact when it comes to ascertaining what should be done. Generally speaking, a person has a special responsibility for taking upon herself the costs of resolving a dilemma if she knowingly brought it about and it would not exist if not for what she did. This may be called the *principle of responsibility*. It does not enjoy universal support but has strong *prima facie* plausibility.[7]

As far as the federation's dilemma is concerned, the principle of responsibility speaks against (ii). This alternative places the whole cost of restoring fishery resources on states that did not contribute to their overexploitation. But the priority principle supports (ii). We may assume, more specifically, that inhabitants of the coast state will die, perhaps in large numbers, unless it is allowed a substantial quota. How are we to balance backward- and forward-looking considerations? In my view, there is an evident duty on the part of other people in F to bail the coast state out. The principle of responsibility is, in other words, outweighed by the priority principle.

It may be objected that this conclusion is misleading because the previous account of the federation's dilemma glosses over two problems. First, even a policy that meets with virtually no opposition is opposed by some citizens, and it seems unreasonable that the latter should be held responsible for their fellows' folly. Second, even if the policy in question is subject to complete consensus, there are children in the coast state who do not share the responsibility for what the adult population does. This arguably suffices to justify the choice of alternative (ii), and there is no need to say anything about the relative force of the principle of responsibility and the priority principle.

In reply to this objection I reiterate my conclusion that the federation has a duty of seeing that the coast state does not suffer severe hardships regardless of whether or not the latter has inhabitants who do not share in the responsibility for depleting fishery resources. Considerations of responsibility and interest-based considerations pull in opposite directions, but it is, all things considered, imperative not to pursue a policy that derives someone of the means of meeting vital needs. I can offer no theory to explain why the priority principle outweighs the principle of responsibility. My conclusion is best described as an intuition of principle. No doubt, one can think of situations in which it comes under considerable strain, but I would be hard put ever to give it up. Most people presumably take the same view.

But what about this case:

> *The other federation's dilemma.* In F', the situation is similar to that in F except for one crucial thing: all member states of F' are highly dependent on incomes from fisheries. This leaves basically two alternatives: either (i) all reduce takes equally for an extended period, or (ii) the coast state is barred from fishing while the other states are allowed annual catches as large as regeneration of fishery resources permits. What should be done?

The other federation's dilemma is a hard case. Whatever way one resolves it, urgent interests will go unmet. This is to say that the priority principle favours neither option, and it may seem that the principle of responsibility tips the balance in favour of (ii). If there is no way of avoiding some people lacking the means of meeting vital needs, those, if any, who knowingly contributed to creating this situation have the weakest claim to protection.

Assume, however, that the coast state is much more populous than all the other states of F' put together. Then there may be a basis for choosing (i) on the ground that this alternative puts a smaller number of lives at stake. So, at any rate, the argument about hard cases in chapter 3 suggests: when lives will be lost anyway, the fewer the better. How far can this argument go to rebut the principle of responsibility? I profess no strong opinion on that score, only an inclination to believe that, if numbers differ considerably, backward-looking considerations will not hold sway.

Notes

1 The assertion is quoted from Walzer (1988: 145), whose presentation of Camus' ideas on moral matters captures the tensions of a dual outlook quite well.

2 It should be noted that this formula is suggested by Shelly Kagan in the course of an effort to throw doubt on the validity of importing personal considerations into normative deliberation. Kagan's own views on the matter will be touched on in chapter 5.

3 *Pace* Jon Elster (1978: 51), who asserts that it may make good sense to accept something as a political goal although it cannot be brought about. Thus, in 'the newly liberated countries where political development is more

advanced than the state of the economy ... it may be a political necessity to proclaim rates of growth or to promise levels of consumption that are plainly impossible at that stage'.

4 Some might think that ordinary desires also place constraints on normative argument because a failure to respect them violates democratic values. If, in particular, a majority of the electorate care less about future needs than present welfare, public policy should reflect this collective preference or else it would be authoritarian. This suggestion misconstrues both democratic values and the nature of normative argument. Its shortcomings are lucidly exposed by Derek Parfit (1983: 32): 'We should distinguish two questions: (a) As a community, may we use a social discount rate? Are we morally justified in being less concerned about the more remote effects of our social policies? (b) If most of our community answer "yes" to question (a), ought our government to override this majority view? The Argument from Democracy applies only to question (b). To question (a), which is our concern, it is irrelevant.... A democrat believes in certain constitutional arrangements. These provide his answer to question (b). How could his commitment to democracy give him an answer to question (a)? Only if he assumes that what the majority wants, or believes to be right, *must* be right. But few democrats do assume this.'

5 The remainder of this section draws heavily on Malnes (1995).

6 Perhaps the greatest power of persuasion lies in the ability to draw on people's personal experience. Thus, 'Bismarck once said that fools learn by experience, wise men learn by other people's experience. By this definition, few men are wise. Before the Bay of Pigs, President Kennedy knew that expert advice should not be accepted without scrutiny, but it took his personal involvement with that disaster to drive this lesson home' (Jervis 1976: 239).

7 John Rawls (1971: 310–15) comes close to claiming that people are never, strictly speaking, responsible for what they do. The weaknesses of his argument are exposed by George Sher (1987: ch. 2).

5
The interests of future people

Introduction

Remote environmental risk cuts across generations in the sense that those who engage in a risky activity will be gone and new people have come into being when adversities eventually materialize. Does the priority principle cut across generations, too? Is it relevant to the assessment and adjudication of environmental dilemmas involving intergenerational conflict of interest?

The conventional view of what we may call inter-temporal morality is a biased one. The interests of contemporaries are widely accorded an urgency that does not attach equally to the requirements of future people. Consider, for example, the political debate about ways of preserving the ozone layer, which protects the earth from ultraviolet radiation. Owing to the risk that it will be destroyed through, *inter alia*, the emission of chlorofluorocarbon (CFC) gases from refrigerators and other electric articles, environmentalists demand that current techniques for the production of these articles be replaced by new, more costly ones. But many, especially developing countries, dismiss such demands. Thus, it is reported that China, 'like countries already industrialized, ... will risk damaging the environment in its drive for prosperity.... After a decade of economic reforms in China, the level of development in many areas is quickly rising toward the stage where consumption of CFCs takes off' (*Christian Science Monitor*, 23–29 March 1989). Even philosophers often start from the idea that the interests of contemporaries come before those of future people. Thus, J. J. C. Smart (1975: 63) speaks of 'localized benevolence', which finds expression in an 'ethics of the present day and generation'. Similarly, Robert Goodin (1985: 169) talks about the 'conventional moral wisdom', which 'inflate[s] the claims of those nearer to us in time and ... reduce[s] correspondingly the claims of those further from us'. In this chapter, I shall first ask whether we are justified in taking such an attitude to posterity if the interests of present and future people are assessed from a purely impersonal point of view. Then I probe other rationales of an 'ethics of the present day and generation', notably the principle of reasonable partiality that was invoked in chapter 4.

The chapter begins by challenging the idea that contemporary welfare comes before the interests of future people, arguing that the priority principle cuts across generations. But the priority principle does not necessarily have the same implications with respect to inter- and intragenerational conflict of interest. New elements come into the picture when the temporal scope of moral assessment widens – considerations that do not detract from the normative force of future people's interests, but still, all things considered, may give reason to downgrade these in comparison with the interests of contemporaries. Several such considerations will be discussed; some stand up to scrutiny, while others do not.

Under 'The presumption of progress', it is argued that the moral importance of meeting future needs should be reduced, because of our ignorance about the remote effects of current policies. The next section asks if such reductions are also warranted on the ground that present people may compensate future ones for the harmful effects of risky policies. This idea is rejected, as is the notion that we may float some duties to posterity because our generation is burdened with a disproportionate part of environmental preservation.

The last section turns to the reality check: should we, for the sake of realism in moral assessment, tilt the balance towards the present when interests of different temporal location are adjudicated? I shall argue that realist considerations warrant no such conclusion. There are ways of allaying the risk that the application of the priority principle to conflict between generations becomes an exercise in utopian thinking.

The priority principle and future people

Before answering the question of whether or not future people come within the scope of the priority principle, it may be useful to start with an easier one: does the priority principle bear on the relationship between people of different societies? Consider, in this connection, the international division of labour. It is arguable that the current division benefits industrialized states at the expense of economic growth in developing countries producing raw materials, and that the former consolidate their favourable position through restrictive trade practices. If so, and granted that the situation of many poor countries is serious enough to put subsistence at stake, are the trade policies of industrialized states at odds with the priority principle? I believe they are, which is to say that the principle bears on the relationship between people of different societies, not just compatriots. The underlying comparison of human interests in terms of normative force makes no mention of nearness or distance in space between those whose subsistence is under threat and those who do the threatening. One does not have to look beyond the physical state of someone who lacks, say, adequate nutrition to find the source of moral concern about this situation. Where the person lives is immaterial in this respect. This does not mean, of course, that neighbours or compatriots may not have duties to one another which they do not have towards people who live outside their vicinities. Neither, however, does the moral importance of such relationships detract from the inherent normative force of vital needs.

Questions about the temporal scope of the priority principle are more tangled. While needy people in distant places are people of flesh and blood, those whose needs may go unmet, say, 200 years from now are nonexistent people. We do not know them, and cannot be sure that they will share our physical and mental constitution; nor can we know for certain that they will ever exist. In view of these uncertainties, is there any reason to abstain from doing things that risk disrupting their potential lives?

If this question is answered in the affirmative, two ancillary problems must be sorted out. (i) Does the duty to posterity cover our relations to future compatriots only or future people everywhere? (ii) Does it extend just to immediately succeeding generations or is its temporal range indefinite? I believe that (i) is redundant. Insofar as a duty to posterity exists, its spatial scope must coincide with the spatial scope of the corresponding duty to contemporaries. If, in other words, the priority principle really cuts across generations, there is no reason why its impact area should vary through time. We may therefore turn from (i) to (ii) and ask whether or not the priority principle bears on how people behave towards only their immediate successors or all succeeding generations. There is no reason for restricting it to the former. If any person who is yet to be born comes within the scope of the principle, all future people do so, because there is no difference between later generations as far as the normative force of their needs is concerned. None need anything today, but all may eventually do so. To be sure, the more remote the generation a person will belong to if she ever exists, the greater the chance that she does not share our physical constitution and, hence, our needs, but granted that she does, meeting her needs is as important as meeting the needs of someone belonging to a more proximate generation. Thus, the two ancillary questions bring us back to the primary one: does our relationship to future people fall within the scope of the priority principle?

I shall answer by way of a simple thought experiment. Suppose a person, P, is to choose between two actions, X and Y, that may adversely affect the vital needs of two nonidentical groups of people, living at different points of time. Doing X will harm some of P's contemporaries, while doing Y risks harming members of some future generation. Here is the story: P owns a factory whose refuse products include toxic liquids that may be disposed of by either emptying them into a nearby lake – which is alternative X – or pouring them into containers and sinking these in the lake – which is alternative Y. If the latter course is taken, the liquids will be safely stored as long as the containers are corrosion resistant, which they are for at least 200 years. After that, however, rust may destroy the containers and release their contents. Let us assume, moreover, that alternative X immediately causes serious health problems to people who live by the lake. Alternative Y leaves contemporary residents unaffected, but represents a risk to the wellbeing of their descendants several generations hence. If X were the only alternative available, the priority principle would call for closing down P's factory. P's economic interests in the further running of his factory would be outweighed by other people's vital needs. This follows from the ranking of needs and desires according to normative force. But P may also, as indicated, store the toxic liquids in containers and pass on the risk involved to posterity. Does this open an opportunity for P to avoid closing down his factory? Will the existence of alternative Y save it?

There are two reasons why one may think that doing Y is morally permitted while doing X is not. First, the victims of Y are possible people only, not real ones. They do not exist at the time P contemplates his choice. Thus, there are no people of whom we may say that doing Y has adverse effects on *them*. But does this really matter? As long as there will be victims of Y, there is reason for moral concern. There is no need to know anything about those deprived of subsistence in order to understand the harmful nature of such deprivation. Identities appear irrelevant on this score.

Second, there is reason to worry about prospective victims of Y only if there will actually be people around at the time its adverse effects materialize. Thus, a reason not to worry is that future people may never come into being. Not only are their identities unknown, but their existence cannot be taken for granted. In view of this, it may be argued: (i) the abstract possibility of death and suffering should not deter anyone – it is imperative only to abstain from activities that may in fact have adverse effects; (ii) P knows for sure that no one will live by the lake a hundred years from now; hence, (iii) P has no reason to be concerned about his containers, which will preserve their contents twice as long. Granted the premises of this argument, the conclusion follows. But obvious problems pertaining to premise (ii) imply that this line of reasoning is unsound. We do not, of course, know the whereabouts of future people, nor, for that matter, the length of human existence. (There are things, like nuclear war, which may cut it short.) Chances are, however, that countless generations remain and that patterns of settlement will stay fairly stable. In place of premise (ii) we may at most invoke an assumption to the effect that P is *uncertain* whether or not there will be any future victims of his doing Y. But then (iii) no longer follows. The mere possibility that future people will not exist does not justify a total disregard of their subsistence.

If P nevertheless has no qualms about disposing of toxic liquids in a way that may affect the health of future residents in the lake area, he is akin to the person who, in Adam Smith's (1759/1976: 102–3) parable, threw 'a large stone over a wall into a public street ... without regarding where it was likely to fall'. He showed an 'insolent contempt of the happiness and safety of others,' exposing them 'to what no man in his sense would chuse to expose himself'. To be sure, 'Nothing would appear more shocking to our natural sense of equity, than to bring a man to the scaffold merely for having thrown a stone carelessly into the street without hurting any body. The folly and inhumanity of his conduct ... would in this case be the same; but still our sentiments would be very different.' As far as P is concerned, we will not be around to adjust our moral sentiments upon learning whether or not he actually harms anyone by storing toxic liquids in the lake. All we can observe is his current insolent contempt for the safety of others – a glaring piece of inhumanity today even if it proves to be harmless. It shows that the relationship between members of separate generations falls within the scope of the priority principle. I shall call this conclusion the *rejection of moral indifference to posterity*.

One might ask if a stronger conclusion is warranted – a conclusion to the effect that the vital needs of future people are morally on a par with the vital needs of contemporaries. Those who answer in the affirmative subscribe to what we may call the *fundamental principle of nondiscrimination*. It is the idea that future people's lives and health are just as important as the lives and health of present people, and the most urgent interests of posterity count for more than the lightweight interests of contemporaries. The fact that a present person needs something here and now, while people who will populate the earth at some future point in time have only potential needs, is in itself no reason to be more concerned about today's needs than tomorrow's. If there is no other reason to accord priority to living persons, a conflict pitting their interests against the interests of some future generation will have to be dealt with as if the two groups were contemporaries. As long as we attend only to the urgency of interests, there is no legitimate basis for discrimination.

The rejection of moral indifference to posterity is not obviously at odds with discrimination in favour of present people. Even the fundamental principle of nondiscrimination does not necessarily run counter to discrimination in practice. To contend that the needs of present and future people are on a par *qua* needs is not tantamount to saying that they are on a par *all things considered*. The fundamental principle of nondiscrimination is, in other words, weaker than an *unconditional principle of nondiscrimination*.

Granted the fundamental principle, it is an open question whether or not the unconditional principle holds true. It depends on what reasons there are to deflate the normative force of future needs and pay more regard to contemporaries than posterity, the timeless importance of vital needs notwithstanding. There may, in other words, be noninterest-based considerations that permit discrimination in favour of present people. These considerations will not imply that the interests of a living person have greater normative force than the interests of an unborn one simply because the former exists here and now while the latter has yet to be born. It rather starts from the idea that present and future needs are on a par, but concludes that, all things considered, existing people merit greater concern than posterity.

To get an idea of how this can come about, consider the following argument. Whatever the intrinsic importance of meeting future people's needs, they stake claims on present people only insofar as current choices make a tangible difference for future wellbeing. Otherwise, it is of little concern today how posterity fares. Thus, if we knew that manna would fall from heaven to feed our descendants, we could disregard their nutritional requirements. Or if, more realistically, new and ample deposits of energy were certain to be discovered within a century, there would be room for discounting future energy needs when deciding on current levels of consumption. This is not to say that the expectation of such developments detracts from the intrinsic importance of future needs. It does not attenuate their normative force, but if future people will subsist anyway, or if, conversely, they will suffer deprivations that none of our efforts could possibly allay, then it would be irrational on our part to sacrifice anything for their sake. Here lies a viable reason why the fundamental principle of nondiscrimination may be compatible with the repudiation of conclusive impartiality.

In the remainder of this chapter, I shall look at both good and bad reasons for discounting the future. Neither the theory of interest nor the priority principle will be questioned, but they are supplemented by arguments that drive a wedge between the rejection of indifference to posterity and the far stronger position expressed by the unconditional principle of nondiscrimination. The timeless scope of the priority principle notwithstanding, it turns out that meeting present people's needs is a more urgent concern than meeting the needs of future generations.

The presumption of progress

I shall start by looking at a nonmoral reason for discounting the interests of future people in normative assessments of environmental policy. It comes from J. J. C. Smart – a utilitarian philosopher who, at the fundamental level, dismisses all forms of discrimination. He appeals to the sentiment of 'generalized benevolence, that is, the disposition to seek happiness, or at any rate, in some sense or other, good consequences for all mankind, or perhaps for all sentient beings' (Smart 1975: 7). This implies that one should always act so as to bring about the best possible overall consequences, taking into account everyone who will be affected – immediately or eventually, directly or indirectly – by what is done. The welfare of each person, alive or yet to be born, counts for exactly the same. The only basis for treating people differently is the strength of interests, and present people enjoy no special concern or protection. Smart's utilitarianism is, in other words, tantamount to embracing the fundamental principle of nondiscrimination.

There may, however, be nonmoral reasons for favouring present people under particular circumstances. Indeed, Smart ends up rejecting the unconditional principle of nondiscrimination. Before we come to that, it will pay to consider a case where such circumstances do not obtain and Smart sees no reason to safeguard the interests of contemporaries before those of posterity.

> If it were known to be true, as a question of fact, that measures which caused misery and death to tens of millions today *would* result in saving from great misery and from death hundreds of millions in the future, and if this were the only way in which it could be done, then it *would* be right to cause these necessary atrocities.
>
> (Smart 1975: 63)

The argument involves four premises. First, it presupposes the fundamental principle. Vital needs are assumed to be of equal concern regardless of where and when their fulfilment is jeopardized. Second, the case Smart conjures up pits interests of equal urgency against one another. Whatever we do – whether or not certain measures are taken – someone will be deprived of subsistence. Third, both alternatives – sacrificing future lives to present ones or vice versa – are certain to cause misery and death. There is, accordingly, no basis for discounting the bad consequences of either alternative on account of the chance that they will not ensue after all. Fourth, the number of lives that will be lost or wrecked in the future if the measures in question are not taken, is greater than the number of present lives which will suffer the same fate if the measures are renounced. Granted these premises, Smart resolves the conflict by counting the people who may meet with death or misery now and later. As future lives are worth no less than present ones, he accepts that saving a greater number in the long run justifies a smaller, albeit substantial, toll today. This line of argument echoes an assumption of chapter 3: if two mutually exclusive actions are both certain to make some people incapable of meeting vital needs, the fact that one action will have far more victims that the other is a reason to choose the other.

A standard objection to this kind of grisly calculus turns on the very structure of utilitarian morality. Utilitarianism leaves no room for deontological duties to forego certain actions – actions that are inherently and unconditionally wrong. Murder – the deliberate killing of people who do not themselves threaten other people's lives – is the most obvious candidate for an act that is *tout à fait* wrong and absolutely prohibited from a moral point of view. But even if deontological duties exist and rule out political measures which involve outright murder, they may not bar the type of measures Smart has in mind. Take the case of a government that can eradicate grave poverty and feed thousands in the short run only by utilizing arable land so intensively that the soil will become barren and eventually unusable for food production, leaving a larger future population without enough to eat. Note that the analogy of murder is no longer appropriate, as there is no question of deliberately killing anyone for the sake of posterity. Is, then, the government justified in saving contemporary lives through measures that lead to more deaths through hunger and undernourishment among later generations? Or should it, as Smart suggests, adopt the policy that saves the greatest number of lives over time? Some might argue that other countries are not justified in letting such things happen. They must assist the unfortunate government in feeding its present citizens and so eliminate the stark choice between contemporary and future lives. This is surely a pertinent observation, and I shall later indicate that positive duties beyond borders are indeed derivable from negative duties to posterity. The problem is, however, that external assistance may fail to materialize in the world as we know it, and the stark choice easily becomes a real dilemma which cannot be evaded. The problem therefore remains: is the government justified in sacrificing present lives for the sake of future ones? If we assume that (i) the fundamental principle of nondiscrimination is a valid position, (ii) lives will be lost whatever we do, and (iii) there will be a much larger toll if present lives are saved than if they are sacrificed, then (iv) sacrificing present lives is the least evil. This may be called the *stark* argument.

As indicated above, Smart has more to say about the kind of case under consideration. His embrace of the fundamental principle notwithstanding, he moves away from the unconditional principle by acknowledging that the vital need of a present person should under certain circumstances be accorded greater weight than a future person's need. On second thoughts, the case for sacrificing present lives is based on 'confusing probabilities with certainties' (Smart 1975: 63–4).

> One thing we should now know about the future is that large-scale predictions are impossible. Could Jeremy Bentham or Karl Marx (to take two very different political theorists) have foreseen the atom bomb? Could they have foreseen automation? Can we foresee the technology of the next century? Where the future is so dim a man must be mad who would sacrifice the present in a big way for the sake of it.
>
> (*Ibid.*: 64)

Smart's final argument involves no revision of his basic moral premises. It is, he avows, 'empirical facts, and empirical facts only, which will lead the utilitarian to say this' (*ibid.*). The facts in question are historical evidence that economic and technological developments, as well as political events, often take quite unexpected turns. Because we can never know for certain what the future effects of contemporary actions will be, premise (ii) of the stark argument must be withdrawn. It is a casualty of what chapter 1 called the *presumption of progress* – the idea that, given time, human beings have a considerable propensity for developing means to cope with emerging threats to their welfare. With this presumption comes a justified hope that future people will eventually be saved from great misery and death even if the only measures that currently seem capable of saving them were to be renounced.

The presumption of progress relates to the fact that ignorance about the future effects of environmental policies has two faces – one malign and one benign. The malign face shows when risky actions or policies actually engender adverse outcomes – perhaps more adverse than foreseen today. The benign face emerges when ominous warnings prove unwarranted – when, that is, potential adversities fail to materialize and more agreeable outcomes come about instead.

Two issues may serve to illustrate this distinction. The malign face of ignorance is associated with the risk that ecological change generates threshold effects. Small and ostensibly harmless disruptions of natural processes may accumulate into substantial damage, as when pollution reaches a level where it can no longer be absorbed by rivers or lakes, giving rise to a rapid deterioration of water quality. The benign face of ignorance relates, for example, to technological innovation, which is known to take place all the time, history being full of successful efforts to overcome adversities that seemed insurmountable until some novel technique emerged. Either face of ignorance provides a reason why it is wrong to sacrifice contemporary lives with a view to saving future ones. The moral costs of such measures are gross and certain to ensue; the gains, albeit great, are only probable and must be reduced. Sacrifices for the sake of posterity may, in particular, prove redundant on account of unforeseen, advantageous developments. It is conceivable, for example, that future generations will have the benefit of agricultural technology that renders current problems of food production inconsequential. This is what the presumption of progress is about. It gives reason to say, as Smart does, that a person must be mad who would sacrifice the present in a big way for the sake of the future.

Underlying Smart's final argument is a principle of rational choice under ignorance, which was considered in chapter 3 – the *disaster avoidance principle*. If we are to choose between two actions, A_1 and A_2, and if (i) O_i is the worst possible outcome of A_1, (ii) O_j is the worst outcome of A_2, (iii) and O_i is somewhat better than O_j, (iv) and the probability that O_i will result from A_1 is markedly higher than the probability that O_j will result from A_2, then (v) A_2 is the preferable alternative. In the case under consideration, there are two alternatives – either adopting drastic measures (X) or not adopting such measures (Y). The worst possible outcome of X – misery and death to tens of millions today – is better than the worst outcome of Y – misery and death to hundreds of millions in the future. Granted the presumption of progress, however, the likelihood that Y will produce its worst outcome is smaller than the likelihood that X will do so. Hence, it is rational to choose Y and run the risk of the very worst outcome in order to minimize the chance that any outcome as bad as the worst possible outcomes of X and Y will ensue.

It may be objected that one of the assumptions behind the disaster avoidance principle is not fulfilled in the case discussed by Smart.

The principle presupposes that the worst possible outcomes of the alternatives we compare do not differ greatly in terms of value. Thus, the risk we run by minimizing the chance of adversity must not be a risk of something very much worse than what may take place if we rather choose the action whose worst outcome is best. But it is arguable that the misery and death of hundreds of millions of people is very much worse than the misery and death of tens of millions. In view of this, we should perhaps choose X so as to maximize the minimum value – which is what the maximin principle recommends – rather than choosing Y with a view to minimizing the risk of adversity.

There is no need to settle this matter here. Smart is right that doing things that bring deaths and miseries to tens of millions of present people is to sacrifice the present in a big way, but the future death and misery of hundred of millions are a far bigger sacrifice. It would be grotesque, however, to attempt anything like a precise estimate of proportionality when dealing with such moral horrors. While numbers count when it comes to deciding what we ought to do in case vital needs will go unmet whatever we do, there is no linear relationship between the number of people who come to harm and the evil of their situation.

Yet, Smart's grisly example teaches one definite lesson. If the vital needs of contemporary people are sacrificed, a disastrous outcome takes place. Disaster may also take place if future needs are put at stake, but there is also a chance that, owing to things beyond our control, it will not. Unforeseen events may, as it were, come to the rescue of future people and avert the risks we impose on them. Hence, a policy that gives precedence to contemporary people minimizes the risk that vital needs will go unmet. To be sure, we are at a loss to estimate the likelihood that developments take a benign turn, and for all we know they may not do so, but history testifies to the benign face of ignorance and justifies some measure of trust in the future.[1]

Smart's argument implies that the fundamental principle of nondiscrimination is not just compatible with, but may naturally lead to, a deviation from nondiscrimination in practice when it comes to dealing with remote environmental risk. What we have found is a nonmoral reason for discrimination in favour of present people. This will have to be justified by empirical and, thus, contingent facts, and if there is nothing supporting the presumption of progress, the argument withers. For all practical purposes, however, it amounts to a rejection of the unconditional principle of nondiscrimination. Ignorance will inevitably surround the effects of current policies on the future standard of living, and the benign face of ignorance, while it may not always show, can never be counted out. This is an incontrovertible truth about historical development.

Compensation

The distance in time between perpetrators and possible victims of remote environmental risk makes it possible for the former to compensate the latter for future adversities. It is arguable, in general, that a duty not to inflict risk on others is annulled if potential victims can be sure of full compensation should they actually come to harm. This may be called the *straightforward* compensation argument. (A more roundabout argument comes later.) It permits me, for example, to play football in the courtyard provided I guarantee that any broken window will be promptly replaced at my expense.

Two problems lurk. First, compensatory measures are rarely instantly effective. There is usually a time-lag before adverse effects can be amended. Second, the exposure to risk may in itself be a torment which guarantees of compensation cannot fully redeem. If, however, risks affect only future people, both problems are beside the point. There is ample time for devising compensatory measures that will be in place when environmental risks eventually materialize, and future people, being unborn, experience no torment from exposure to the risks in question. Hence, the straightforward compensation argument suggests that remote environmental risk should sometimes, all things considered, be tolerated.

Brian Barry invokes kindred considerations in support of the view that each generation may consume nonrenewable resources despite the concomitant risk to later generations.

> I propose that future generations are owed compensation in other ways for our reducing their access to easily extracted and conveniently located natural resources. In practice, this entails that the combination of improved technology and increased capital investment should be such as to offset the effects of depletion.
>
> (Barry 1983: 17)

'The important thing,' Barry (*ibid.*: 20) says, 'is that we should compensate for the reduction in opportunities brought about by our depleting the supply of natural resources, and that compensation should be defined in terms of productive capital'. But although the straightforward compensation argument may work in this particular case, it is not applicable to all problems of environmental risk. Thus, the potentially harmful effects of pollution and climate change do not easily lend themselves to amendment, as there is little contemporaries can do to offset the attendant reduction in opportunities. While I do not dispute the general validity of the compensation argument, I doubt whether it goes very far towards lifting the moral constraints on risky environmental policy.

There is, however, more to the question of compensation. Besides the possibility that perpetrators of risk may buy exemption from the priority principle by compensating potential victims, it is arguable that they may sometimes be justified in violating the principle unless *they* receive adequate compensation for the costs of abiding by it. Generally speaking, a person who inflicts risk on others through an activity which in itself is morally neutral and will be costly for him to renounce, may be owed an indemnity for renouncing this activity. The absence of such an indemnity weakens, if it does not annul, the force of considerations that tell against his right to engage in the risky activity. This may be called the roundabout compensation argument. Robert Nozick (1974: 81–2) puts it in suggestive terms:

> Some types of action are generally done, play an important role in people's lives, and are not forbidden to a person without seriously disadvantaging him. One principle might run: when an action of this type is forbidden to someone because it *might* cause harm to others and is especially dangerous when it does, then those who forbid in order to gain increased security for themselves must compensate the person forbidden for the disadvantage they place him under. This principle is meant to cover forbidding the epileptic to drive while excluding cases of involuntary Russian roulette.... The idea is to focus on important activities done by almost all.... Almost everyone drives a car, whereas playing Russian roulette ... is not a normal part of almost everyone's life.

This is to say that the priority principle may leave a moral remainder. If the activity it forbids in a given case does not inflict risk gratuitously, but only as an unintended side-effect, the agent may require something in return for observing the principle. The case for compensation seems particularly pertinent if the prohibition is discriminatory and – as in the instance of epileptics – directs itself to people who cannot be faulted for doing things more dangerously than others. Placing them at a disadvantage because of their bad luck is unfair, although the reasons why they ought to renounce a dangerous activity may be strong and beyond doubt.

Who should offer compensation in such a case? Two groups may come into question. There is first, as Nozick indicates, the people who stand to gain from others' constraint. It is not, I think, that these beneficiaries incur debts of gratitude; after all, the behaviour in question exposes them to great risk. It is rather a question of solidarity – sharing the extra costs that happen to fall on certain people and might just as well have fallen on others. Second, it seems that compensation should come from those, if any, who are responsible for the unfortunate fact that some people cannot perform ordinary activities without creating high risks. As far as epileptics are concerned, no one can be taken to charge for the dangerous nature of their driving, but responsibility can be traced in other cases.

This is the case, for example, with respect to many risks of environmental destruction that the activities of present people give rise to. They happen to live at a time when processes of environmental decline are reaching dangerous levels. Halting them so that future needs can be met may require substantial reductions of economic growth and the standard of living. If, however, previous generations – especially the most recent ones – had shown more restraint, lighter demands would have fallen on present people. As things are, contemporaries may be obliged to sustain a lower standard of living than their immediate predecessors owing to their unfortunate temporal location. This is unfair, and contemporaries seem entitled to some compensation for their disproportionate sacrifices. Not compensating them leaves a moral remainder. This argument basically accords with the principle of responsibility that was invoked in chapter 4.

However, neither previous people nor future ones are capable of compensating contemporaries for current sacrifices. If the priority principle is to be observed, unfairness inevitably ensues. Which is worse, then, violating the principle or imposing unfair costs on those who engage in risky behaviour? I believe this conflict between perpetrators and victims of remote environmental risk must be resolved in favour of the victims. The unfortunate fact that present people are left with disproportionately large costs does not outweigh our concern about future needs or wipe out the attendant duties to posterity. Unmet vital needs are a greater evil than unfairness.

Yet, the special problems of poor, developing countries call for a more subtle conclusion. They are particularly disadvantaged by the fact that the priority principle places disproportionate demands on members of the present generation. Economic affluence has just come within their reach, and sacrifices for the sake of future people are liable to place new obstacles in the way of sustained growth. It is therefore pertinent to ask if developing countries in particular should be compensated for such sacrifices by those who benefit from their restraint or those, if any, who are responsible for placing a disproportionate burden of sacrifice at their feet. The beneficiaries of present restraint will be future people who cannot reciprocate, but there are some people alive today who should be accorded duties of compensation on grounds of responsibility. Citizens of affluent countries have long pursued policies that overtax the environmental basis of global development. If more restraint had been shown on their part, there would have been greater room for welfare improvements in developing countries. To be sure, some of the greatest damages were effected by previous generations of affluent people who can no longer be called to account. It is not unreasonable, however, to hold their descendants – the present citizens of rich states – vicariously responsible. They are the ones who currently benefit from previous excesses, and they are amply endowed with the means of financing compensatory measures. Hence, considerations of fairness make it a duty of affluent countries to transfer resources to poor countries so as to enable them to attain affluence without compromising future people's vital needs.

Realism and remote risk

Normative assessment should be imbued with a spirit of realism, or so I argued in chapter 4. We shall see what this implies as far as the application of the priority principle to problems of remote environmental risk is concerned.

Assume that an activity poses a significant risk to the subsistence of future people while its renunciation would only frustrate lightweight contemporary interests. According to the priority principle, this activity should be renounced. The realism of this suggestion is, however, questionable. Can present people be expected to sacrifice much – or anything at all – for the sake of posterity? Will demands for the reduction of contemporary standards of living be heeded when intended beneficiaries are future people?

It will be convenient to start from some arguments of *Our Common Future*, the report of the World Commission on Environment and Development (WCED 1987). Like this book, *Our Common Future* takes the fulfilment of vital – or, as it usually says, basic – needs to be the highest moral priority. What it recommends under the head of 'sustainable development' is, broadly speaking, a process of economic growth that is not at the expense of subsistence anywhere or any time. To be sure, social welfare involves more than the fulfilment of vital needs. People also 'have legitimate aspirations for an improved quality of life' (*ibid.*: 43), but their realization is legitimate only under certain conditions.

> Living standards that go beyond the basic minimum are sustainable only if consumption standards everywhere have regard for long-term sustainability.... Meeting essential needs depends in part on achieving full growth potential, and sustainable development clearly requires economic growth in places where such needs are not being met. Elsewhere, it can be consistent with economic growth, provided the content of growth reflects the broad principles of sustainability and non-exploitation of others.
>
> (*Ibid.*: 44)

Disregarding exploitation, what is implied by saying that the content of economic growth should reflect 'the broad principles of sustainability'? On a natural reading, it implies that those whose vital needs have already been met should advance their welfare only by means which do not compromise other people's needs, now or later. But *Our Common Future* goes on to hint at another, more lenient interpretation of the principle of sustainability. It says that sustainable development 'requires the promotion of values that encourage consumption standards that are within the bounds of the ecologically possible and to which all can reasonably aspire' (*ibid.*). This implies, in particular, that if current values do not dispose contemporary people to lower their economic aspirations out of regard for future people, new values should be fostered, but contemporaries are not held to a stern demand of lowering their aspirations right away.

The tension between a stern and a lenient interpretation of sustainability is suggested in the following comment on how current economic endeavours may affect the future supply of natural resources:

> With minerals and fossil fuels, the rate of depletion and the emphasis on recycling and economy of use should be calibrated to ensure that the resource does not run out before acceptable substitutes are available. Sustainable development requires that the rate of depletion of non-renewable resources should foreclose as few future options as possible.
>
> (*Ibid.*: 46)

The first sentence corresponds to a stern interpretation of sustainability. It tells us in no uncertain terms to ensure that nonrenewable resources are not consumed at a rate that leads to shortages before adequate substitutes have been found. But the second sentence retracts, prescribing a level of consumption that puts *as few constraints as possible* in the way of future efforts to meet the needs that nonrenewable resources currently serve.

This raises the question of where the limits of possibility should be drawn. What future options are present people capable of not closing and which cannot possibly be kept open? The lenient interpretation of sustainability suggests that it ultimately depends on people's values. Their contents determine the limits of possibility at any time. Yet, values are malleable, and efforts should be made to alter attitudes that make people neglectful of the interests of posterity. There is no reason not to confront them with the 'impartial truth' – to recall Godwin's phrase – about the greater force of future needs than present desires, but some reason to expect that this truth will gradually get a grip on them.

The ideas just ascribed to the World Commission on Environment and Development accord well with what was said about realism and the principle of reasonable partiality in chapter 4. That discussion also, however, underlined the risk of utopianism: making demands on people that put such strains on the common and natural course of their passions that morality as such is brought into disrepute. What can be done to allay this risk in the present context? In chapter 4, I alluded to one way of altering basic preferences which might seem relevant.[2] One may try the indirect route of inducing people to think anew about their values by influencing their beliefs about what the values in question really imply. Perhaps people's readiness to care for future generations can be influenced by this indirect route. Let me start with an easier question: can we affect their propensity to care for their own future by working on their beliefs? There are some who stare fixedly at possible future adversities and others who do not care what is in store for them. Such dispositions have the appearance of psychological givens, not in the sense that they never change, but by virtue of being to some extent inexplicable. On second thoughts, however, there is obviously room for argument about attitudes to time. Myopia and foresight are not completely irrational dispositions. To a person who is absorbed by worries about future misfortunes, one may rightly point out the irrationality of undue anxiety about things that, as Hume (1740/1985: 428) says, are better left 'to the care of chance and fortune'. And to someone who brushes aside the risk that things may go wrong, one can explain how badly she will fare if worst comes to worst. If a person does 'not regard you' when you talk him 'of his condition thirty years hence' (*ibid.*: 428–9), the reason may be that he does not fully appreciate what you say. Present experiences are tangible and rich in details, but images of the future tend to be pale. This implies that myopia may be corrected by graphic descriptions of future adversities.

> First, although we typically realize that our future interests generate *some* reason for our acting, we often fail to assess the weight of these reasons accurately (and thus fail to act prudently, not realizing in particular cases that this is supported by the balance of reasons). A more vivid understanding of how one's future interests can be affected in a given case can lead one to better assess the proper weight of the reasons generated by the fact. Second, and perhaps more importantly, in other cases we may well already judge that the balance of reasons favor acting prudently, but we may not be *moved* by this realization. Vividness helps here too: we are better able to give reasons for their proper influence on our behavior when we have vivid beliefs.
>
> (Kagan 1989: 287)

This suggests a way of instilling foresight in someone who is disposed to myopia. Does the same strategy lend itself to efforts to persuade people not to care more about their own lightweight interests than future people's needs? At the least, it will allow us to eliminate the risk that appeals to the priority principle in intergenerational conflicts are seen as an ethics of fantasy. Even if such appeals are resisted and sometimes fall on deaf ears, as they undoubtedly will, they hardly become branded as utopianism if accompanied by vivid accounts of the grisly facts of life-threatening environmental destruction.

I shall wind up this section by briefly mentioning a different route to dispelling the lack of realism that lurks in calls for an enhanced concern about future people's needs. One sometimes gets the impression that implementing duties to posterity is all a question of large-scale reforms of institutions and policies, as only such reforms will make a significant difference to the future standard of living. It is not demanded that individuals reform their daily lives with a view to eliminating practices that tax the environment, as individual contributions to environmental decline are too trivial to matter. Thus, the quantities of carbon dioxide which private cars at idling speed emit to the atmosphere are insignificant compared with the emissions from large industries or motorways. Substantial reductions of pollution will take place only through political regulations that, among other things, limit factory effluent and restrict motoring. Whatever private concessions individuals undertake on their own are quite inconsequential and of no avail.

Such considerations apparently underlie Robert Goodin's (1985: 163–4) view of the institutional strategy as the only viable one when it comes to discharging intergenerational – as well as international – duties.

> we do have a responsibility to aid distant peoples in distress but ... it is a collective rather than an individual responsibility.... Saying that the responsibility is a collective one does not exempt individuals from responsibility; it merely changes the character of their responsibilities. Individual members of the collectivities in question have a responsibility to cooperate in whatever schemes are organized to discharge collective responsibilities.... They also have a responsibility to undertake political action – at the national and/or international level, depending upon where they can be most efficacious – that is designed to ensure that such schemes are in fact organized.

This assertion turns on duties between countries in particular, but is carried over to the relationship between generations:

> In tracing the implications of this argument for responsibilities to future generations, little more needs to be said than has already been said in connection with the previous applications to ... foreign aid. Responsibilities toward future generations should also be seen primarily as collective ones.
>
> (*Ibid.*: 178)

This may be correct as far as it goes, but it leaves out something important. Large-scale reforms need popular support to be carried into effect. It is doubtful whether such support is elicited most effectively by moral campaigns on behalf of the same reforms. Rather, new and potentially demanding standards may be taken to heart only if they are inculcated through small-scale modifications of individual behaviour. The crucial thing is to cultivate moral virtues – to bring out 'what may be called *good dispositions* – that is, some readiness on occasion voluntarily to do desirable things which not all human beings are just naturally disposed to do anyway, and similarly not to do damaging things' (Warnock 1971: 76). It is, at any rate, a plausible assumption that close causal relationships exist between private dispositions and political attitudes, and it is therefore unwise to neglect small-scale morality in efforts to bring about large-scale reform.

Notes

1 The lesson from Smart is echoed in the following argument by Charles Beitz (1981: 342): 'While I cannot advance a general theory of moral choice under uncertainty, it seems clear that any plausible theory would include a principle roughly like the following: given a choice among strategies for serving distinct sets of equally urgent needs, we should choose the mixture of strategies that offers the greatest probability of success. Such a principle, I suggest, explains why concern for the welfare of future generations should not be allowed to undercut efforts to assure members of the present generation a decent minimum level of satisfaction of its urgent needs.' Beitz does not mention numbers and so evades the hard question that was just discussed.

2 This paragraph draws on Malnes (1995).

6

Conclusion

The argument in retrospect

An environmental dilemma exists if one group of people may see their standard of living reduced on account of an activity that damages the natural environment, while another group may suffer hardships if steps are taken to avoid the damage in question. Either way, someone's interests will fulfilled to a lesser extent than they could have been. This may happen to present people as well as members of future generations.

I have argued that it is more important that vital needs are met than that desires for an even higher standard of living are satisfied. This assumption rests on an impersonal assessment of two states of affairs: the notion that it is worse, objectively speaking, to be without adequate nutrition, health care or protection from the elements than lacking, say, a varied diet or a commodious place to live. The priority principle was derived from this assumption and says that it is imperative not to engage in activities whose impact on the natural environment deprives or risks depriving someone of the means of meeting vital needs, unless the risk is too small to be a reasonable source of concern.

The last clause of the priority principle alludes to the fact that activities which risk damaging the environment may represent greater or lesser threats to people's lives and health. Insofar as risks can be estimated numerically, a critical level – say 10% – must be fixed, and no higher chance that vital needs will go unmet should be tolerated. If, moreover, the number of people whose lives may be affected is large, the critical level should be lowered. When only ordinal estimates of risk are available, a risk is significant if the likelihood that needs will go unmet exceeds the likelihood that they will not or if the two likelihoods are about the same. And, finally, if neither numerical nor ordinal estimates can be had, any demonstrable risk is significant.

Standards were also proposed for dealing with hard cases: situations in which lives may be lost irrespective of whether threats to the environment are tolerated or eliminated, as well as dilemmas involving no interest as urgent as a vital need. As far as the former are concerned, I argued in chapter 3 that they should be resolved according to the principle of minimizing expected mortality provided alternative solutions can be told apart on this score. If they cannot, moral considerations call for equalizing various people's chances of survival. When it comes to the other kind of hard case – situations in which some people's desires can be fulfilled only by dint of frustrating others' – precedence should be accorded to desires for the preservation of pristine nature in societies where the opportunity of experiencing pristine nature is about to be eliminated.

These normative standards draw their justification from interest-based considerations. But the forward-looking, impersonal aspect of morality that finds expression in such considerations is not the whole of it. One element that comes in addition is backward-looking considerations, summed up in the assumption that a person has a special responsibility for taking upon herself the costs of resolving a dilemma if she knowingly brought it into being and it would not exist if not for what she did. Another additional element is personal concerns that justify people in according greater importance to the interests of their nearest and dearest than objective assessments warrant. The two counterweights to interest-based considerations were elaborated in chapter 4, which also, more tentatively, hinted at ways of combining divergent perspectives to reach conclusions about what morality demands, all things considered, in particular situations.

As I said above, people who may suffer from what is done in the face of an environmental dilemma can be alive today or belong to a later generation. Impersonally speaking, the fulfilment of a future person's vital needs matters as much as the fulfilment of the needs of present people. But we may still be justified in showing greater concern for the living than posterity, or so I argued in chapter 5. One reason lies in the fact that risks of remote, negative contingencies should often be reduced on the ground that social and technological progress may in due course eliminate the threats involved. Another, more problematic reason is that present people can be justified in displaying partiality towards those of their own ranks.

How does the argument of this book compare with competing perspectives on environmental policy? As normative problems in this field are currently the subject of an ardent philosophical interest that makes itself felt in a spate of publications, comparisons must necessarily be selective. I shall consider two challenges. One, coming from ecological ethics, says that nature should be preserved for its own sake, irrespective of how its degradation or preservation affects human life conditions. The other, attributed to 'green' parties and movements, implies that the instrumental conception of nature as a basis of material development is too narrow. Both challenges will be rejected.

Against ecological egalitarianism

This book adopts an anthropocentric perspective. It locates the value of the natural environment in its instrumental role as a means of meeting human needs and gratifying desires. This is the way problems of environmental policy are usually approached, but another perspective is sometimes adopted. Here is Paul Kennedy (1993: 99–100) reflecting on the disappearance of tropical forests in Latin America and seeing several reasons for concern:

> The first is the destruction of the way of life of many innocent tribes. It is also the case that these forests have the world's greatest store of plant and animal species by far ... and the destruction of this fantastic array of biodiversity would deal a heavy blow to humankind's constant need to keep renewing (and improving) pest-resistant and productive crops. Population pressure leading to deforestation would thus curb global agriculture's ability to renew itself – and to provide for the additional billions of consumers. It would also be a blow to the fecundity and fascination of life itself.

The last sentence suggests that tropical forests should be preserved because biodiversity is valuable as such. It corrects the anthropocentric bent of Kennedy's argument, but at the same time renders it more opaque. While the rationale of improving pest-resistant and productive crops is immediately comprehensible, this hardly goes for the idea that forests must be preserved out of concern for 'the fecundity and fascination of life itself'. The anthropocentric view of nature has a basis in common sense that the nonanthropocentric view lacks. Best, then, to begin with the idea that nature should be conserved for the sake of human beings, but now it is time to confront the nonanthropocentric perspective.

What can we make of the idea that environmental risks are a source of concern in their own right, quite apart from our apprehension their social and economic implications? The nonanthropocentric view of nature is espoused by proponents of *ecological ethics*. I shall consider their claim to have transcended 'human chauvinism' (Routley and Routley 1979), but there is no room for a comprehensive examination of the many versions of ecological ethics that have emerged in recent years. I aim, by engaging a select sample of articulate representatives, only to show that any concern about the environment rooted in a nonanthropocentric view of nature must take second place, all in all, to a concern about vital human needs. In whatever way the view that nature matters for its own sake is interpreted, it will not constitute a core element of the theory of value that is to guide environmental policy.

Rather than discussing one particular work, I shall focus on Roderick Frazier Nash's (1989) succinct and sympathetic summary of two seminal contributions – Arne Naess' and George Sessions' extensive writings on 'deep ecology'.[1] Their basic assumption is that *Homo sapiens* does not stand out from other living beings as far as moral status is concerned. Human self-realization is neither more nor less important than the self-realization of animals and plants and rivers, and so on. 'The central idea [is] the right of every form of life to function normally in the ecosystem.... Rivers [have] a right to be (or function as) rivers, mountains to be mountains, wolves to be wolves, and ... humans to be humans' (Nash 1989: 146–7). Some aspects of normal functioning are, however, more important than the rest. They are vital or basic, while others are peripheral or excessive, and the former come before the latter, irrespective of whose functioning we are talking about. This thesis of ecological egalitarianism marks a radical departure from conventional thought. It calls for 'the end of dualism', that is, for humans to 'step back into the life community as a member and not the master' (*ibid*.: 148).

> Naess and Sessions could explain in 1984 that environmental impact, even killing, was ethically acceptable as long as it was done 'to satisfy *vital*

> needs.' The antipode of 'vital' or 'basic' in the minds of deep ecologists was 'peripheral,' 'excessive,' or 'nonvital.' The central liability of modern technological civilization was that it had lost the ability to distinguish between these antipodes. In unmodified ecosystems, as would be found under wilderness conditions, predator and prey existed in a balance of vital needs. Hierarchy, domination, exploitation, and power – all hated words in the deep-ecology lexicon – did not exist in nature.... Antelope kill grass, lions kill antelope, flies or pre-civilized humans kill lions – all in the course of a series of what Naess would call self-realizations.... But technological humans, unlike their own hunting and gathering ancestors, possessed the power to alter ecosystems beyond their vital or legitimate need to survive.
> (*Ibid.*: 147–8)

The egalitarian thesis presupposes a distinction between vital and peripheral aspects of normal functioning. This is most reasonably drawn in parallel with the previous distinction between different human interests. Thus, every being, not just humans, has vital needs that relate to conditions whose fulfilment is a prerequisite of its continued existence and normal functioning. Just as a person needs food and protection from the elements to survive, a tree dies if it is felled, and a mountain ceases to exist unless it is protected from blasting operations. Naess and Sessions invoke an array of interests that can be attributed, more or less metaphorically, to animals, plants, landscapes, and so on. Moreover, every being may be advantaged in ways that go beyond the necessary conditions of its normal functioning – a person by the gratification of desires, a tree by clean air, and a mountain by the prevention of all traffic on it. This line of argument points towards the following preliminary statement of ecological egalitarianism: people may engage in activities that alter the natural environment if and only if these activities do not interfere with vital aspects of anyone's or anything's functioning.

The thesis of ecological egalitarianism challenges the anthropocentric view of nature at its root. The distinctive status of human interests as a source of moral concern is called into question. True, much remains to be sorted out before ecological egalitarianism becomes a determinate guideline for environmental policy. Its preliminary statement does not tell us what to do if nonhuman nature has to be destroyed in order that human needs be met, nor demarcate the beginning and the end of the 'life community', nor indicate the vital functions of every member of this community. (What are the prerequisites of, say, a river being what is? Its free flow in every direction where it will flow if left to itself?) None of these problems need detain us here, however. We may concentrate on the more basic question of whether or not ecological egalitarianism is the proper perspective on environmental matters.

The distinctive status that human interests are accorded on the conventional, anthropocentric view has two aspects. In the first place, human desires generally command greater normative force than the vital needs of animals. Not, of course, that they always do. The gratuitous infliction of pain on any sentient being is unconditionally wrong, however fun some deranged individuals may find it. It is widely and reasonably thought, however, that cattle may be raised for food despite the easy availability of vegetarian nourishment. In the second place, a basic premise of anthropocentric thinking is that nonsentient beings do not warrant moral concern in their own right. Most people can, like W. K. Frankena (1979: 11), see

> no reason, from the moral point of view, why we should respect something that is alive but has no conscious sentiency and so can experience no pleasure or pain, joy or suffering.... Why, if leaves and trees have no capacity to feel pleasure or to suffer, should I tear no leaf from a tree? Why should I respect its location any more than a stone in my driveway, if no benefit or harm comes to any person or sentient being by moving it?

The thesis of ecological egalitarianism rejects both aspects of the distinctive status conventionally accorded to human interests. It calls for a fundamental transformation of the way we think about environmental matters.

Should we heed the call? However alluring it may be made to sound, its utopian character undermines it. As things stand, there is no chance that humans will put their interests on a par with the interests of animals and nonsentient beings. If, therefore, normative problems are to be approached in the spirit of realism, the perspective of ecological ethics must be set aside. It may, accordingly, be rejected on the grounds laid out chapter 2: indisputable facts about the limits of human motivation. There is no practical point in demanding, for example, that people stop cutting down trees for the sake of trees themselves. Doing so presupposes an unreal aptitude for self-transcendence. In general, the idea that the same normative force attaches to the interests of trees, mountains and human beings is light years away from what Hume calls the 'common measures of duty'.

To be sure, the spirit of realism involves no capitulation before these measures. As I said in the previous section, letting the personal point of view into moral thinking is not tantamount to throwing out impersonal concerns, only conceding that some revisions of common measures are too demanding to constitute serious options for most people. Which are and which are not may be hard to tell, but not when it comes to the thesis of ecological egalitarianism. It is motivationally inaccessible to almost everyone.

Against the green theory of value

The *green theory* of value, espoused and defended by Robert E. Goodin (1992), owes its name to the claim that it constitutes the ideological basis of the green parties that emerged in Western Europe during the 1980s. Whether Goodin actually succeeds in eliciting the philosophical underpinnings of these parties need not concern us here. The green theory is of interest anyway, standing midway between the conventional approach to environmental problems and proposals for an altogether new environmental ethics.

The green theory is at one with conventional thinking in assessing environmental policies by reference to the place of the natural environment in human life. Yet, it parts company with convention in depicting nature as much more than a mere means for the sustenance of physical needs and wants. According to the green theory, the natural environment rather takes on the appearance of a medium of meaningful human lives. This viewpoint is summed up in three theses (Goodin 1992: 37):

(i) People want to see some sense and pattern to their lives.
(ii) That requires, in turn, that their lives be set in some larger context.
(iii) The products of natural processes, untouched as they are by human hands, provides precisely that desired context.

Thesis (i) is an empirical postulate which Goodin takes to be an indisputable truth about human psychology: 'What makes people's lives seem valuable to those who are living them is the unity and coherence of the projects comprising them' (*ibid.*: 38). This assumption borders on vacuity if 'unity and coherence' can consist in just any 'sense and pattern' – however haphazard – that may be attributed to a life. But thesis (ii) deflects the lurking vacuity of (i) with a vengeance. It implies that people want a very distinct coherence to their lives – 'to see some continuity between their inner worlds and the external world,' and 'to lead their lives in harmony with the external world' (*ibid.*: 38). This is to say that some ways of life definitely do not reflect the requisite quest for coherence. Thus, continuity between inner and outer worlds was no concern of the Manichees of the third century, who tried to sever themselves from every aspect of the physical world, which they saw as the embodiment of evil. It is likewise no concern of modern techno-freaks who envision an existence among artificial objects in 'virtual reality'. The addition of the second thesis implies, in other words, that the first thesis no longer states an inescapable empirical truth. When (i) is seen in conjunction with (ii), it refers to a substantive conception of what lends sense to life – not an idiosyncratic conception, to be sure, but a distinct one.

In thesis (iii), Goodin distances the green theory further from commonplace psychology. He assumes that the allegedly universal aim of continuity between one's own life and the external world is best achieved by placing oneself in the context of forces that operate 'independently' of, and often 'in spite of', human activities (*ibid.*: 40). He asserts, moreover, that nature provides this context better than any other medium. Two questions may be raised about thesis (iii). First, nature is not the only independent context that a life can be set in. What about history, including the immensely complex and uncoordinated process through which human beings have harnessed and transformed nature? The answer is that history is, after all, less independent of human design than natural processes. 'People who live more in harmony with nature – in traditional English villages, rather than postmodern Los Angeles – are living more in a context that is outside of themselves, individually or even collectively' (*ibid.*: 51). This leads to the second question about thesis (iii): do people – even those who find coherence in a harmony between their inner lives and the external world – really crave for a context that is maximally independent of themselves? I doubt that many do. Anyway, it should be clear by now that the green theory of value, far from being neutral in questions of good and bad, rests on controversial assumptions about what makes life meaningful. In essence, it calls for the conservation of untouched nature as an especially propitious basis for a life that displays an especially valuable form of unity and coherence. There is nothing trivial or self-evident about such a theory of value.

I shall not proceed by questioning the green theory on conceptual or philosophical grounds, although it stands in urgent need of both elaboration and justification. Let us instead assume, for argument's sake, that many people really want what Goodin has in mind for them – to be enmeshed in natural processes that operate outside and partly in spite of themselves. Is the existence of this desire a good reason not to interfere with nature by, say, decimating forests, destroying animal species or causing climate change? It goes for any desire that its fulfilment is a good thing as long as more weighty moral considerations do not call for its renunciation. If, however, one person's desire can be fulfilled only in ways that make another incapable of meeting vital needs, it has to be renounced unless this is an unrealistic demand. So, at any rate, the priority principle and the proviso of realism imply. We are adjudicating between interests of different normative force – desires on one side and needs on the other – and the want of living one's life in a certain way belongs to the first category, together with some desires alluded to in chapter 2: scholars' desire for well equipped libraries and everyone's desire for a higher net income. To be sure, the desire for a particular way of life is principle dependent in the sense that it cannot be described without reference to what it is reasonable for a person to want (Rawls 1993: 82–3). For this reason, it may impress us more strongly than frivolous desires for material goods and sensual pleasures. As far as normative force is concerned, however, one person's survival and normal biological functioning override another's quest for continuity between an inner world and the natural environment. The green theory of value might, of course, be put forth as an alternative to the theory of interest, involving an altogether different conception of the relative normative force of vital needs and certain desires. But Goodin offers no argument to buttress such a conception, and it is highly counterintuitive. One person's opportunity of living in a context of untouched nature does not count for more than another's survival.

A final look at climate policy: what should be done?

The argument of chapters 2 through to 5 equips us to look anew at the problem described in detail in chapter 1: what should be done in response to the risk of global warming and climate change? This is an environmental dilemma in that both options – taking comprehensive measures aimed at curbing emissions of carbon dioxide or refraining from such measures – may adversely affect someone's standard of living. And the choice will have to be made under ignorance, as the outcomes of taking or not taking precautionary measures against an alleged risk are not known for sure. We have imperfect knowledge about (i) what, if any, changes the global climate is currently undergoing, (ii) the effects of climate change on living conditions in various parts of the world, and (iii) the economic and social effects of comprehensive measures aimed at halting such change.

As to effects on living conditions, we saw in chapter 1 that unabated emissions of carbon dioxide and other greenhouse gases are most likely to precipitate climate change that in turn results in dwindling supplies of drinking water and decreasing agricultural yields, and otherwise deprives people of what they need to preserve life and health. Granted that an activity raises a significant risk of depriving someone of the means of subsistence if its likelihood of doing so is greater than the likelihood that it does not, the priority principle implies that anthropogenic emissions of greenhouse gases ought to be abated. But two things might speak against comprehensive abatement measures: (i) the association of an equally serious risk to life and health with reductions of anthropogenic greenhouse gas emissions, or (ii) qualms about the lack of realism in demands for reductions: do they go beyond the limits of reasonable partiality?

As far as (i) is concerned, we already know from chapter 1 that eliminating up to one-fifth of existing carbon dioxide emissions is unlikely to hurt the world economy very much, but economic adversity will most likely result if emissions are more than halved. This is to say that comprehensive measures of the kind proposed by the Intergovernmental Panel on Climate Change (IPCC) render the risk of economic decline higher than the chance of continued growth.

The next question is whether or not the risk associated with reductions of greenhouse gas emissions is as serious as the risk of climate change. We have only ordinal probability estimates to go on and cannot compare the likelihoods of the most adverse outcomes of either reducing or not reducing emissions. But are the adversities themselves comparable? Climate change will, at worst, leave people without the means of meeting vital needs in many parts of the world.

May reductions in greenhouse gas emissions, largely obtained through decreasing use of fossil fuels, do the same? It seems clear that no serious adversity can result from modest reductions of around 20%, but the 60% cutback proposed by the IPCC is another matter. It will be recalled that William Nordhaus, a leading economist, sees 'depression' in the wake of such a cut, while others anticipate 'massive economic dislocation' (Solow 1991: 26).

Does economic dislocation amount to loss of subsistence? In other words, does the choice of climate policy answer to the description of a hard case in which every alternative may deprive someone of the means of meeting vital needs? In industrialized countries, economic depression is plainly no threat to lives and health. It may force people to live less expensively, but will not come near to taking away what they need to survive. The average level of affluence attained in Europe, North America and Oceania provides a far more robust protection from poverty than a 2% reduction in gross national product can undo.

The situation of developing countries is different from that of industrialized states. As the former lack 'the technology to increase conservation and produce energy from nuclear and renewable resources,' it will 'prove all but impossible' for them to more than halve greenhouse gas emissions while taking major steps towards improving nutrition standards, health care and education (Lave 1991: 15–16). The traditional road to development, which involves extensive energy consumption and high tolerance of pollution, will be blocked. At best, developing countries will be capable of maintaining the status quo despite substantial restrictions on energy use, but status quo implies a life on the poverty line (Mitchell 1992: 5–6).

There is, accordingly, no realism in demanding that the peoples of these countries restrict their use of fossil fuels. The question of whether or not they should take drastic steps to curb greenhouse gas emissions must be answered in the negative on both scores indicated above: the steps they might take to avert the risk of climate change are bound up with their incurring an equally serious economic risk, and there is no realism in urging them to incur such a risk.

It may be objected that the inhabitants of developing countries will suffer no serious hardship if people in the affluent part of the world compensate them for the cost of redirecting energy use away from fossil fuels, as they ought to do. This argument presupposes,

References

Allison, Graham. 1971. *Essence of Decision*, Boston, Little, Brown and Company.

Aquinas, Thomas. 1967. *Thomas Aquinas,* Utvalg ved Knut Erik Tranøy (excerpts by Knut Erik Tranøy), Oslo, Pax.

Arrow, Kenneth J. 1966. Exposition of the theory of choice under uncertainty, *Synthese*, 16:4, 253–69.

Balling, Robert C. 1992. *The Heated Debate: Greenhouse Predictions Versus Climate Reality*, San Francisco, Pacific Research Institute.

Barry, Brian. 1983. Intergenerational justice in energy policy, in Douglas MacLean and Peter G. Brown (eds): *Energy and the Future*, Totowa, Rowman & Littleheld.

Barry, Brian. 1989. *Theories of Justice*, London, Harvester-Wheatsheaf.

Beitz, Charles R. 1981. Economic rights and distributive justice in developing societies, *World Politics*, 33:3, 321–46.

Blum, Lawrence. 1988. Gilligan and Kohlberg: implications for moral theory, *Ethics*, 98:3, 472–91.

Brown, Peter G. 1992. Climate change and the planetary trust, *Energy Policy*, 20:3, 208–22.

Cameron, James and Abouchar, July. 1991. The precautionary principle: a fundamental principle of law and policy for the protection of the global environment, *Boston College International and Comparative Law Review*, 14:1, 1–28.

Charlson, R. J., Schwartz, S. E., Hales, J. M., Cess, R. D., Coakley, J. A. Jr, Hansen, J. E. and Hofmann, D. J. 1992. Climate forcing by anthropogenic aerosols, *Science*, 255, 24 January, 423–30.

Charlson, Robert J. and Wigley, Tom M. L. 1994. Sulfate aerosol and climatic change, *Scientific American*, February, 28–35.

Cline, William R. 1992. *The Economics of Global Warming*, Washington, DC, Institute for International Economics.

Cline, William R. 1993. *Costs and Benefits of Greenhouse Abatement: A Guide to Policy Analysis*, Paris, Organization for Economic Co-operation and Development and International Energy Agency (OCDE/GD(93)90).

Colglazier, E. William. 1991. Scientific uncertainty, public policy, and global warming: how sure is sure enough? *Policy Studies Journal*, 19:2, 61–72.

Cox, Loren C. 1991. Introduction: facts and uncertainties, *The Energy Journal*, 12:1, 1–8.

Descartes, Rene. 1628/1967. Rules for the direction of the mind, in *The Philosophical Works of Descartes, Volume I*, Cambridge, Cambridge University Press.

Easterbrook, Gregg. 1992. A house of cards, *Newsweek*, 1 June, 20–5.

Ellsaesser, Hugh W. 1992. The threat of greenhouse warming is maintained by ignoring much of what we know, in S. Fred Singer (ed.): *The Greenhouse Debate Continued: An Analysis and Critique of the IPCC Climate Assessment*, San Francisco, The Science and Environmental Policy Project.

Elster, Jon. 1978. *Logic and Society*, Chichester, John Wiley.

Elster, Jon. 1983. *Explaining Technical Change*, Cambridge, Cambridge University Press.

Elster, Jon. 1988. Introduction, in J. Elster and R. Slagstad (eds): *Constitutionalism and Democracy*, Cambridge, Cambridge University Press.

Elster, Jon. 1989. *Solomonic Judgements*, Cambridge, Cambridge University Press.

Firor, John. 1990. *The Changing Atmosphere*, New Haven, Yale University Press.

Fishkin, James S. 1982. *The Limits of Obligation*, New Haven, Yale University Press.

Føllesdal, Dagfinn. 1979. Some ethical aspects of recombinant DNA research, *Social Science Information*, 18:3, 401–19.

Frankena, W. K. 1979. Ethics and the environment, in K. E. Goodpaster and K. M. Sayre (eds): *Ethics and the Problems of the 21st Century*, Notre Dame, University of Notre Dame Press.

Gerholm, Tor Ragnar. 1992. Sustainable scenarios? An assessment of IPCC's CO_2 emission assumptions, in S. Fred Singer (ed.). *The Greenhouse Debate Continued: An Analysis and Critique of the IPCC Climate Assessment*, San Francisco, The Science and Environmental Policy Project.

Glover, Jonathan. 1977. *Causing Death and Saving Lives*, Harmondsworth, Penguin Books.

Godwin, William. 1798/1976. *Enquiry Concerning Political Justice*, Harmondsworth, Penguin.

Goodin, Robert E. 1978. Uncertainty as an excuse for cheating our children: the case of nuclear waste, *Policy Sciences*, 10:1, 25–43.

Goodin, Robert E. 1985. *Protecting the Vulnerable*, Chicago, University of Chicago Press.

Goodin, Robert E. 1992. *Green Political Theory*, Cambridge, Polity Press.

Gray, Vincent R. 1992. The IPCC report on climate change: an appraisal, in S. Fred Singer (ed.): *The Greenhouse Debate Continued: An Analysis and Critique of the IPCC Climate Assessment*, San Francisco, The Science and Environmental Policy Project.

Griffin, James. 1986. *Well-Being. Its Meaning, Measurement and Moral Importance*, Oxford, Clarendon Press.

Grubb, Michael. 1989. *The Greenhouse Effect: Negotiating Targets*, London, The Royal Institute of International Affairs.

Hampshire, Stuart. 1989. *Innocence and Experience*, London, Allen Lane.

Harris, John. 1974. The Marxist conception of violence, *Philosophy and Public Affairs*, 3:4, 192–220.

Harsanyi, John C. 1975. Can the maximin principle serve as a basis for morality? A critique of John Rawls' theory, *American Political Science Review*, 69:2, 594–606.

Hayes, Peter. 1993. North–South carbon abatement costs, in Peter Hayes and Kirk Smith (eds): *The Global Greenhouse Regime. Who Pays?*, Tokyo, United Nations University Press.

Hempel, Lamont C. 1993. Greenhouse warming: the changing climate in science and politics, *Political Research Quarterly*, 46:1, 213–39.

Herzog, Don. 1989. *Happy Slaves. A Critique of Consent Theory*, Chicago, University of Chicago Press.

Hey, Ellen. 1991. The precautionary approach. Implications of the revision of the Oslo and Paris Conventions, *Marine Policy*, July, 244–54.

Howarth, R. B. and Monahan, P. A. 1992. *Economics, Ethics, and Climate Policy*, Berkeley, Lawrence Berkeley Laboratory, University of California.

Hume, David. 1740/1985. *A Treatise of Human Nature*, Oxford, Clarendon Press.

Humphreys, Paul. 1989. *The Chances of Explanation*, Princeton, Princeton University Press.

IPCC (Intergovernmental Panel on Climate Change). 1990. *Climate Change: The IPCC Scientific Assessment*, Cambridge, Cambridge University Press.

IPCC (Intergovernmental Panel on Climate Change). 1992. *1992 IPCC Supplement*, WMO/UNEP

Isaksen, Ivar. 1992. *The Climate Issue, Present State of Knowledge*, Policy Note 1992, 4. Oslo, Center for International Climate and Energy Research.

Isaksen, Ivar S. A. and Fuglestvedt, Jan S. 1993. Uenighet og usikkerhet knyttet til klimaspørsmål (Disagreement and uncertainty regarding climate issues), *Cicerone*, 2:1, 4–6.

Jackson, Frank. 1991. Decision-theoretic consequentialism and the nearest and dearest objection, *Ethics* 101:3, 461–82.

Jervis, Robert. 1976. *Perception and Misperception in International Politics*, Princeton, Princeton University Press.

Jervis, Robert. 1988. Realism, game-theory, and cooperation, *World Politics*, 40:3, 317–49.

Kagan, Shelly. 1984. Does consequentialism demand too much? Recent work on the limits of obligation, *Philosophy and Public Affairs*, 13:3, 239–54.

Kagan, Shelly. 1989. *The Limits of Morality*, Oxford, Clarendon Press.

Kamm, F. M. 1993. *Morality, Mortality. Volume I*, New York, Oxford University Press.

Kavka, Gregory S. 1987. *Moral Paradoxes of Nuclear Deterrence*, Cambridge, Cambridge University Press.

Kennedy, Paul. 1993. *Preparing for the Twenty-First Century*, New York, Random House.

Kevles, Daniel J. 1989. Paradise lost, *The New York Review of Books*, 21 December, 32–8.

Kiehl, J. T. and Briegleb, B. P. 1993. The relative roles of sulfate aerosols and greenhouse gases in climate forcing, *Science*, 260, 16 April, 311–14.

Laslett, Peter and Fishkin, James S. 1992. Introduction: processional justice, in Peter Laslett and James S. Fishkin (eds): *Justice Between Age Groups and Generations*, New Haven, Yale University Press.

Lave, Lester B. 1991. Formulating greenhouse policies in a sea of uncertainty, *The Energy Journal*, 12:1, 9–21.

Leggett, Jeremy. 1992. Global warming: the worst case, *Bulletin of Atomic Scientists*, June, 28–32.

Locke, John. 1690/1984. *Two Treatises of Government*, London, Dent.

Luce, R. Duncan and Howard Raiffa. 1957. *Games and Decisions*, New York, John Wiley.

Lunde, Leiv. 1991. *Science or Politics in the Global Greenhouse?* Lysaker, The Fridtjof Nansen Institute.

Mackie, J. L. 1977. *Inventing Right and Wrong*, Harmondsworth, Penguin.

Malnes, Raino. 1992. Philosophical argument and political practice: on the methodology of normative theory, *Scandinavian Political Studies*, 15:2, 117–34.

Malnes, Raino. 1995. 'Leader' and 'entrepreneur' in international negotiations. A conceptual analysis, *European Journal of International Relations*, 1:1, 87–112.

Mintzer, Irving M. 1992. Living in a warming world, in Irving M. Mintzer (ed.): *Confronting Climate Change*, Cambridge, Cambridge University Press.

Mitchell, James K. and Ericksen, Neil J. 1992. Effects of climate change on weather-related disasters, in Irving M. Mintzer (ed.): *Confronting Climate Change*, Cambridge, Cambridge University Press.

Mitchell, Sir William. 1992. Reflections on global climate change, in S. Fred Singer (ed.): *The Greenhouse Debate Continued: An Analysis and Critique of the IPCC Climate Assessment*, San Francisco, The Science and Environmental Policy Project.

Morgan, M. Granger. 1993. Risk analysis and management, *Scientific American*, July, 32–41.

Morrisette, Peter M. 1991. The Montreal Protocol: lessons for formulating policies for global warming, *Policy Studies Journal*, 19:2, 152–61.

Mueller, John. 1988. The essential irrelevance of nuclear weapons, *International Security*, 13:2, 55–90.

Naess, Arne. 1987. *Ecology, Community and Lifestyle: Ecosophy T*, Cambridge, Cambridge University Press.

Nagel, Thomas. 1986. *The View from Nowhere*, New York, Oxford University Press.

Nagel, Thomas. 1991. *Equality and Partiality*, New York, Oxford University Press.

Nash, Roderick Frazier. 1989. *The Rights of Nature*, Madison, University of Wisconsin Press.

Nordhaus, William D. 1990. Count before you leap, *Economist*, 7 July, 19–22.

Nordhaus, William D. 1991a. Economic approaches to global warming, in Rudiger Dornbusch and James M. Poterba (eds): *Global Warming. Economic Policy Responses*, Cambridge, The MIT Press.

Nordhaus, William D. 1991b. The cost of slowing climate change: a survey, *The Energy Journal*, 12:1, 37–66.

Nozick, Robert. 1974. *Anarchy, State, and Utopia*, New York, Basic Books.

Oppenheim, Felix. 1987. National interest, rationality, and morality, *Political Theory*, 15:3, 369–89.

Parfit, Derek. 1983. Energy policy and the further future: the social discount rate, in Douglas MacLean and Peter G. Brown (eds): *Energy and the Future*, Totowa, Rowman and Littleheld.

Parry, Martin L. and Swaminathan, M. S. 1992. Effects of climate change on food production, in Irving M. Mintzer (ed.): *Confronting Climate Change*, Cambridge, Cambridge University Press.

Rapoport, Anatol. 1980. Various meanings of "rational political decisions", in Leif Lewin and Evert Vedung (eds): *Politics as Rational Action*, Dordrecht, D. Reidel Publishing Company.

Rawls, John. 1971. *A Theory of Justice*, New York, Oxford University Press.

Rawls, John. 1993. *Political Liberalism*, New York, Columbia University Press.

Ross, W. D. 1930. *The Right and the Good*, Oxford, Clarendon Press.

Routley, R. and Routley, V. 1979. Against the inevitability of human chauvinism, in K. E. Goodpaster and K. M. Sayre (eds): *Ethics and the Problems of the 21st Century*, Notre Dame, University of Notre Dame Press.

Samuelson, Robert J. 1992. The end is not at hand, *Newsweek*, 1 June, 32.

Sanders, John T. 1988. Why the numbers should sometimes count, *Philosophy and Public Affairs*, 17:1, 3–14.

Scanlon, T. M. 1975. Preference and urgency, *Journal of Philosophy*, 72:19, 655–70.

Scheffler, Samuel. 1992. *Human Morality*, New York, Oxford University Press.

Schneider, Stephen H. 1989. *Global Warming*, Cambridge, The Lutterworth Press.

Sen, Amartya. 1985. Rights and capabilities, in Ted Honderich (ed.): *Morality and Objectivity*, London, Routledge & Kegan Paul.

Sessions, George. 1987. The deep ecology movement: a review, *Environmental Review*, 11:2, 105–25.

Sher, George. 1987. *Desert*, Princeton, Princeton University Press.

Singer, S. Fred. 1989. Overview, in S. Fred Singer (ed.): *Global Climate Change. Human and Natural Influences*, New York, Paragon House.

Singer, S. Fred. 1992. Executive summary, in S. Fred Singer (ed.): *The Greenhouse Debate Continued: An Analysis and Critique of the IPCC Climate Assessment*, San Francisco, The Science and Environmental Policy Project.

Singer, S. Fred, Revelle, Roger and Starr, Chauncey. 1993. What to do about greenhouse warming: look before you leap, in Richard A. Geyer (ed.): *A Global Warming Forum. Scientific, Economic, and Legal Overview*, Boca Raton, CRC Press.

Sjöberg, Lennart. 1979. Strength of belief and risk, *Policy Sciences*, 11:2, 39–57.

Smart, J. J. C. 1975. An outline of a system of utilitarian ethics, in J. J. C. Smart and Bernard Williams, *Utilitarianism: For and Against*, London, Cambridge University Press.

Smith, Adam. 1759/1976. *The Theory of Moral Sentiments*, Oxford, Clarendon Press.

Snyder, Jack. 1984. *The Ideology of the Offensive*, Itacha, Cornell University Press.

Solow, Andrew R. 1991. Is there a global warming problem? in Rudiger Dornbusch and James M. Poterba (eds): *Global Warming. Economic Policy Responses*, Cambridge, The MIT Press.

Stone, Christopher D. 1993. *The Gnat is Older than Man. Global Environment and Human Agenda*, Princeton, Princeton University Press.

Taurek, John M. 1977. Should the numbers count? *Philosophy and Public Affairs*, 6:4, 293–316.

Walzer, Michael. 1988. *The Company of Critics*, London, Peter Halban Publishers.

Warnock, G. J. 1971. *The Object of Morality*, London, Methuen.

Warrick, Richard A. and Rahman, Atiq A. 1992. Future sea level rise: environmental and socio-political considerations, in Irving M. Mintzer (ed.): *Confronting Climate Change*, Cambridge, Cambridge University Press.

WCED (World Commission on Environment and Development). 1987. *Our Common Future*, Oxford, Oxford University Press.

Williams, Bernard. 1975. A critique of utilitarianism, in J. J. C. Smart and Bernard Williams, *Utilitarianism: For and Against*, London, Cambridge University Press.

Wilson, Deborah and Swisher, Joel. 1993. Exploring the gap. Top-down versus bottom-up analyses of the costs of mitigating global warming, *Energy Policy*, 21:3, 249–63.

Index